Modern Telecommunications
Basic Principles and Practices

By
Martin Sibley

CRC Press
Taylor & Francis Group
Boca Raton London New York

CRC Press is an imprint of the
Taylor & Francis Group, an **informa** business

CRC Press
Taylor & Francis Group
6000 Broken Sound Parkway NW, Suite 300
Boca Raton, FL 33487-2742

First issued in paperback 2023

Library of Congress Cataloging-in-Publication Data

Names: Sibley, M. J. N. (Martin J. N.), author.
Title: Modern telecommunications : basic principles and practices / Martin J. Sibley.
Description: Boca Raton : CRC Press, 2018. | Includes bibliographical references and index.
Identifiers: LCCN 2017055201| ISBN 9781138578821 (hardback : alk. paper) | ISBN 9781351263603
Subjects: LCSH: Telecommunication--Textbooks.
Classification: LCC TK5101 .S498 2018 | DDC 621.384--dc23
LC record available at https://lccn.loc.gov/2017055201

ISBN 13: 978-1-032-65311-2 (pbk)
ISBN 13: 978-1-138-57882-1 (hbk)
ISBN 13: 978-1-351-26360-3 (ebk)

DOI: 10.1201/b22525

Contents

Preface

Telecommunications is literally all around us – we are surrounded by electromagnetic waves (radio waves) from many, many sources: TV, radio, mobile phones, Wi-Fi, etc. Our modern society relies on communication as never before; just ask any user of a mobile phone! There is a philosophical question as to whether this new era of communications is a good thing or not; however, what is clear is that society has an ever-increasing demand for bandwidth that is satisfied by some very clever technology which is described in this book. Some text books are written to be dipped into whenever the reader requires knowledge of a particular area; others are written to be read from cover to cover. I wrote this text to take the reader on a journey through the fundamentals of telecommunications and then on to discuss various communications systems. It is difficult to predict the future but one thing for certain is that telecommunications, in all its varied forms, will be at the forefront of the technology. I hope you enjoy reading this book as much as I enjoyed writing it.

I would especially like to thank my wife, Magda, my daughter, Emily, and my family both here and abroad for their invaluable help and encouragement. Thanks also go to the referees for their useful comments, in particular Dr. Karel Sterckx of Shinawatra University, Thailand, for proof reading the text and giving valuable feedback.

What sculpture is to a block of marble, education is to the soul.

Joseph Addison (1672–1719)

Martin J.N. Sibley

About the Author

Martin Sibley has a PhD in preamplifier design for optical receivers from Huddersfield Polytechnic. He started his career in academia in 1986, having spent three years as a post-graduate student and then two years as a British Telecom–funded research fellow. His research work has a strong bias to the practical implementation of research and he has taught communications at all levels since 1986. Currently, Mr. Sibley is a reader in communications at the School of Computing and Engineering, University of Huddersfield. He has authored three books and is a member of the Institution of Engineering and Technology and fellow of the Higher Education Academy.

1 Introduction

1.1 HISTORICAL BACKGROUND

Historically, there are two branches of science that underpin the physics behind electromagnetic fields – electrostatics and magnetism. Both of these have been well known for centuries. The ancient Greeks were familiar with electrostatics from their studies of amber and magnetism was known to the Chinese who invented the magnetic compass. The science of electroconduction was developed around 1800 with the invention of the battery; however, it was Michael Faraday (1791–1867) who made the link between electrostatics and electroconduction – they were the same thing. Now there were two sciences – electric current and magnetism.

It wasn't until 1820 that Hans Christian Oersted demonstrated a link between a magnetic field and a constant current (only direct current [dc] at the time). This was the first indication that electroconduction and magnetism were linked. In 1831, Faraday demonstrated that a changing magnetic field could induce a changing current in a wire. So, we now have a changing magnetic field causing a changing current in a wire and a changing current causing a changing magnetic field. It was James Clerk Maxwell who, in 1865, formalised the work of Faraday and unified electricity and magnetism. This laid the foundation of, among other things, special relativity. A fortunate result of this work was the prediction of electromagnetic waves and that light was an electromagnetic wave.

Oliver Heaviside (1893) adapted the theory presented by Maxwell into the four equations we know today as Maxwell's equations. Heaviside also worked on the telegraph system, predicting that performance can be improved by using loading coils. Following on from Maxwell's prediction, there were several attempts to demonstrate radio wave transmission. However, these were considered to be transmission due to inductive coupling and not to be electromagnetic waves themselves. Credit for using electromagnetic waves goes to Heinrich Rudolf Hertz who, in 1888, demonstrated conclusively that the waves existed. It was Marconi, in 1894, who started work on a commercial radio system and in 1897 he started a radio station on the Isle of Wight in the United Kingdom. He was awarded the Nobel Prize in Physics in 1909. This was the start of commercial broadcasting as we know it.

As a society, we now have radio broadcasting, numerous TV channels, the Internet, voice over Internet protocol (IP), video on demand, text messaging, etc. We are a "wired" or even "wireless" society. It is instructive to see how long this has taken: Maxwell formulated his ideas in 1865 and we are now some 150 years later; the optical fibre that is widely used to carry the Internet and data was developed in 1970; broadcast by satellite started in earnest in 1990; we also have digital radio and TV (1998). The pace of change is very fast and it is a very brave person indeed who would predict what the next 20 years will bring. One thing is certain, the fundamentals will not change and that forms the first part of this chapter.

1.2   REASONS FOR ELECTROMAGNETIC COMMUNICATION

Communications, in the form of electromagnetic waves, are literally all around us – radio, TV, mobile phones, Wi-Fi, satellite, etc. These systems are so common that we often take them for granted. But why do we use electromagnetic waves and why do the signals sometimes drop out so that we lose the telephone link or there is no TV signal?

Let us first look at how we communicate in our everyday lives. As a species, we are equipped with the means to communicate – we talk using our mouths and we hear using our ears. This system of communication is very efficient and is replicated throughout the animal world. So, why have we developed an alternative that uses man-made signals?

One of the problems associated with our innate communication system is that it relies on pressure waves carried by the molecules that make up the air (Figure 1.1). If we are in space, where there is no air, sound does not carry since there are no molecules present. Another problem is one of power. If we talk face to face, not much audio power is required to carry on a conversation. However, in a noisy environment, we have to increase our audio power by raising our voices. Eventually, we have to shout and this places a strain on our vocal chords. This is where electronic amplification comes in.

In order to increase audio power, a microphone can be used to convert sounds into an electrical signal and an amplifier is used to boost the signal (Figure 1.2). A loudspeaker is then used to convert the signal back into a pressure wave. In this way, we can overcome the problem of limited audio power by simply increasing the amplification. There is a major difficulty though. If we wish to broadcast to a large

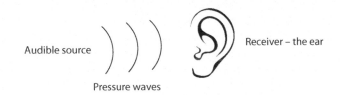

FIGURE 1.1   An audible transmission system.

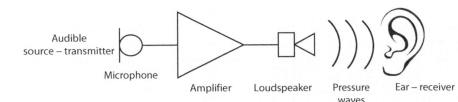

FIGURE 1.2   A powerful audible transmission system.

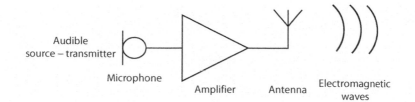

FIGURE 1.3   A basic transmitter.

audience, a city for example, we would require a very large amount of audio power. If the power level is adequate for those far from the loudspeaker, the level for those close to the loudspeaker would be so high that permanent damage to the ear would result. So, there is an obvious limitation if this system is used; but what about electromagnetic waves?

Consider the amplifier system just described. The audible signal is converted into an electrical one by the microphone prior to amplification. What is seldom realised is that an electromagnetic wave is produced after the microphone. It might not be a particularly strong one, but the amplifier boosts it. Instead of the loudspeaker that turns the electrical signal back into an audible one, an antenna (basically a piece of wire) can be used. What we now have is a microphone that converts the audible signal into an electrical one, an amplifier that boosts the electrical signal and an antenna that radiates the electromagnetic signal. This is a very basic transmitter (Figure 1.3). A receiver is required to get the signal back again and this is covered in Section 1.7.

This has solved the problem of limited power to some extent. There is one final problem to be tackled and that is that all the stations occupy the same frequency. This is where modulation comes in.

Let us consider baseband speech from 300 Hz to 3.4 kHz corresponding to the range of a telephone. (Note that this filtering is one reason why music sounds distorted over a telephone. Middle C has a frequency of 261 Hz.) The simple transmitter will generate an electromagnetic wave in this frequency range as it has speech as the input; but what happens if another user wishes to transmit as well? Interference will result as both transmitters are transmitting signals using the same frequency. It is like two people talking at the same time. One solution, which is used today, is to move the baseband signals to different frequencies so that they do not interfere with each other. Each station modulates a particular frequency and different stations transmit on different frequencies. The receiver then tunes into the radio station it requires. This process should be familiar to everyone who listens to the radio and changes station.

Before we move on to the electromagnetic spectrum, we will define some parameters, most notably the frequency and the wavelength.

1.3   SINUSOIDS: SINES AND COSINES

A sinusoid is simply a sine wave or a cosine wave, or a phase-shifted version of either. As a sine wave, it takes the form:

$$v(t) = V \sin \omega t \tag{1.1}$$

where:

    $v(t)$ is the variation in time of the voltage
    V is the peak voltage
    ω is the angular frequency that we will define by Equation 1.3

    A plot of the variation with time is a sine wave, with which most people are familiar. It is instructive to consider a graphical method of generating the sine wave using a technique that uses rotating phasors. Such a scheme is shown in Figure 1.4.

    On the left-hand side of Figure 1.4 is a rotating phasor of length V – the peak amplitude of the sine wave. Its initial position is horizontal and this is time $t=0$. The phasor rotates in a counterclockwise direction about the centre of the circle and completes a revolution in time T seconds. This is the period of the signal. The frequency is how many cycles it makes in one second. The old unit for frequency is cycles per second (cps) while the modern unit is Hertz (Hz). Frequency and period are linked by

$$f = \frac{1}{T} \tag{1.2}$$

    To the right of Figure 1.4 there is a sine wave and this is produced as follows. Consider a phasor at time $t=0$. When observed from right to left, the amplitude appears to be zero. That is the first point on the time plot. A quarter of a cycle later ($t=T/4$), the phasor is vertical and the amplitude of the sine wave is V. At half a cycle ($t=2T/4$), the phasor appears to be zero again. A further quarter cycle gives $t=3T/4$ and the phasor is pointing vertically downwards. Thus, the amplitude appears to be $-V$. At the end of the cycle, $t=4T/4$, the phasor is back at the start and the amplitude is 0. This gives four points on the time plot. To complete the graph, we need to use geometry. On the left-hand side of Figure 1.4, a phasor is drawn between $t=0$ and $t=T/4$. The apparent magnitude of the phasor is $V\sin\varphi$ with $0 \le \varphi \le 360°$ or $0 \le \varphi \le 2\pi$ for a complete rotation. The $\sin\varphi$ term is where our sine wave comes from. In fact, the sine wave is a time plot of the magnitude of the vertical projection of the rotating phasor.

    The frequency is the number of cycles per second and, as each cycle covers 2π radians, we can define an angular frequency, ω, as

$$\omega = 2\pi f \text{ rad/s} \tag{1.3}$$

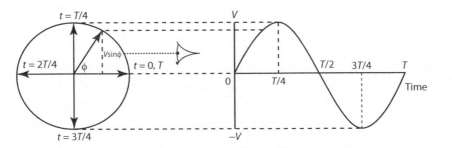

FIGURE 1.4   A phasor diagram representation of a sine wave.

We now have a sine wave given by $V\sin\omega t$ with the ωt term having units of radians. The introduction of a phase shift is easily accomplished by moving the phasor either clockwise, $V\sin(\omega t + \phi)$, or anti-clockwise, $V\sin(\omega t - \phi)$. It should be noted that a sine wave shifted by 90° is a cosine wave.

1.4 THE ELECTROMAGNETIC SPECTRUM: FROM SUBMARINES TO SATELLITES

An important relationship that will be used in the following is one relating frequency and wavelength to the speed of light (approximately 3×10^8 m/s):

$$f = \frac{c}{\lambda} \tag{1.4}$$

where:
  f is the frequency in Hertz
  c is the speed of light in metres per second
  λ is the wavelength in metres

Table 1.1 shows the range of bands in the electromagnetic spectrum. It should be noted that there is no lower or upper limit to the range of frequencies that could be used. The only thing is that there are practical limitations in that the optimum dimension for a dipole antenna is $\lambda/2$ (dealt with later in the book).

TABLE 1.1
The Electromagnetic Spectrum

Name		Frequency Range	Wavelength
Extremely low frequency	ELF	3–30 Hz	10^8–10^7 m
Super low frequency	SLF	30–300 Hz	10^7–10^6 m
Ultra low frequency	ULF	300 Hz to 3 kHz	10^6–10^5 m
Very low frequency	VLF	3–30 kHz	10^5–10^4 m
Low frequency – also known as long wave	LF	30–300 kHz	10–1 km
Medium frequency – also known as medium wave	MF	300 kHz to 3 MHz	1 km to 100 m
High frequency – also known as short wave	HF	3–30 MHz	100–10 m
Very high frequency	VHF	30–300 MHz	10–1 m
Ultra high frequency	UHF	300 MHz to 3 GHz	1 m to 10 cm
Super high frequency	SHF	3–30 GHz	10–1 cm
Extremely high frequency	EHF	30–300 GHz	1 cm to 1 mm
Far infra-red	FIR	300 GHz to 3 THz	1 mm to 100 μm
Mid infra-red	MIR	3–30 THz	100–10 μm
Near infra-red	NIR	30–300 THz	10–1 μm

Before we discuss the various bands and what they are used for, it is instructive to look at the wavelength as quoted in the final column of Table 1.1. As noted previously, a dipole antenna should have a length of $\lambda/2$. If we were to use the extremely low frequency (ELF) band, the antenna size would be five million metres long for 30 Hz. This is clearly impractical. Another problem is that if we were to transmit speech with a range of 300 Hz to 3.4 kHz, we would take up the whole of the ultra low frequency (ULF) band and some of the very low frequency (VLF) band and the antenna length would have to vary as well. Clearly, we can't use very low frequencies for broadcasting audio. Now check what happens as the frequency increases. The antenna size reduces to manageable values and the bandwidth (the difference between the upper and lower frequencies) goes up. Taking the high-frequency (HF) band as an example, the bandwidth is 27 MHz and we can fit many broadcast stations in such a bandwidth.

Now we turn to the usage of the bands.

1.4.1 ELF, Super Low Frequency (SLF), ULF, VLF

The bandwidth of this combination is very low and so we are unable to use them for general broadcasting. However, they do have one very big advantage: very low frequency (LF) signals travel through water, an ability that lessens as frequency increases. It is this characteristic that makes the LF bands suited to communication with submerged submarines. As we can't transmit speech or video because the bandwidth is so small, we must use very slow speed data such as Morse code.

1.4.2 Low Frequency

The LF band is used for broadcasting. Unlike higher frequencies it does not suffer from a large degree of loss and so one or two radio stations can cover a country. A frequency of 87 kHz is used for communication with cavers to a depth of typically 100 m.

1.4.3 Medium Frequency

This used to be the major broadcast band with, for instance, two transmitters covering the whole of the United Kingdom. In view of its coverage, it is a relatively inexpensive way of broadcasting. It is still in use today despite the popularity of very high frequency (VHF) and frequency modulation (FM).

1.4.4 High Frequency

This is a very interesting band. Signals in this band can be reflected off layers in the atmosphere known as the ionosphere. As the name suggests, this layer consists of ionised gases and these act as a reflecting layer. It is this property that enables broadcasters to transmit over very large distances (several 100 km) to other countries.

1.4.5   VERY HIGH FREQUENCY

Signals in this band are quite localised – they are absorbed and scattered by buildings to some extent. This property means that broadcasters are able to broadcast local radio to cities and towns. National broadcasting can be achieved by having a relay system so that the signals are distributed to local VHF stations. As each station provides local coverage, frequency reuse can be used (see Figure 4.32). Digital audio broadcasting (DAB) is broadcast in the region of 225 MHz.

1.4.6   ULTRA HIGH FREQUENCY

Mobile phone systems, discussed in Chapter 6, use this band and there are several frequencies allocated for this: 850, 900, 1800, 1900 and 2100 MHz. UHF signals are very localised with some being absorbed very easily. An example of this is microwave ovens. These operate at a frequency of 2.45 GHz and food is heated by the absorption of the microwave energy. Wi-Fi operates in this range at 2.4 and 5 GHz. The global positioning system (GPS) uses frequencies of 1227.60 and 1575.42 MHz.

1.4.7   SHF

Transmission in this band is by line of sight which is put to very good use by direct broadcast by satellite (DBS). In this system, satellites are placed in geostationary orbit so that they appear at a fixed point above the equator. For broadcast purposes, a parabolic reflector transmitting dish points to the area to be broadcast to. The uplink frequency is approximately 10 GHz and the downlink frequency is approximately 12 GHz.

1.4.8   EHF

This band is currently not used for communications. Instead, it is used for security imaging. The skin is not transparent at such frequencies but clothing is. So, illumination with terahertz (THz) signals will reveal items concealed beneath clothing.

1.4.9   FAR INFRA-RED (FIR), MID INFRA-RED (MIR), NEAR INFRA-RED (NIR)

These bands correspond to light of which visible light is a part. Visible light covers the range 390–700 nm or frequencies 430–770 THz. Our eyes are equipped to receive information in this range. However, if we are talking of communication as we understand radio communication, we can't decode the signals because our eye does not respond fast enough. It is possible to use photodetectors as the receiver and this gives a much faster response. Optical fibre communication uses wavelengths of 850 nm to 1.625 μm.

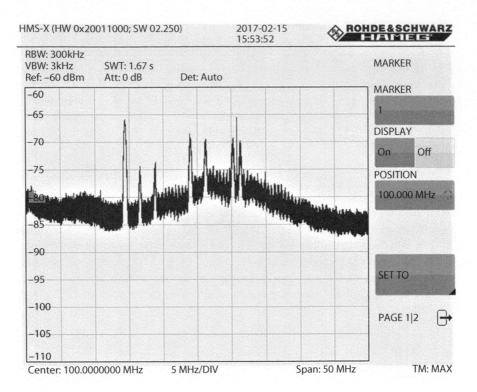

FIGURE 1.5   Measured spectrum of seven radio stations in the FM broadcast band.

1.5   FREQUENCY-DIVISION MULTIPLEXING AND FREQUENCY TRANSLATION

We have just seen that the electromagnetic spectrum can be divided into bands and broadcasting stations can be allocated frequencies in those bands. This is known as frequency-division multiplexing (FDM) and Figure 1.5 shows the spectrum, as measured on a spectrum analyser, of a section of the FM broadcast band. Note that each broadcast station is allocated a particular frequency. As this is transmission at a high frequency, the signals will not travel very far and so it is possible to reuse the same frequencies if the stations are physically remote from one another.

The process of attaching the baseband (speech, music or video) to a HF carrier is known as *modulation*; the various types are dealt with in Chapters 3 and 4. However, there is one very important process that we need to discuss here and that is frequency translation. Modulation takes place at a low frequency and so it is necessary to translate the signal to a higher broadcast frequency. The modulated signal is mixed to a higher frequency in a mixer (Appendix I and Figure 1.6).

In a mixer, the output is the product of the two inputs. If

$$v_1 = V_1 \cos \omega_1 t$$

$$v_2 = V_2 \cos \omega_2 t$$

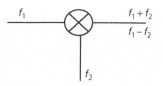

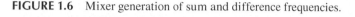

FIGURE 1.6 Mixer generation of sum and difference frequencies.

the output will be the product of these two inputs:

$$\text{Output} = v_1 v_2$$

$$= V_1 \cos \omega_1 t V_2 \cos \omega_2 t \qquad (1.5)$$

$$= \frac{V_1 V_2}{2} \cos(\omega_1 + \omega_2)t + \frac{V_1 V_2}{2} \cos(\omega_1 - \omega_2)t$$

where we have made use of a trig identity (see Appendix II). We normally only want one of the two products and so we have to filter out the unwanted frequency. This can be done using a tuned circuit, which is considered next.

1.6 TUNED CIRCUITS: STATION SELECTION

Tuned circuits are fundamental building blocks in radio systems. They can select particular bands on the electromagnetic spectrum as well as amplify signals and reduce noise. They are simply a parallel or series combination of an inductor (L) and a capacitor (C). The parallel combination (Figure 1.7) is most widely used in telecommunications and so is discussed in detail. (Appendix III applies also.) The inductor and the capacitor react in different ways to changes in frequency. The reactance of the inductor, X_L, increases with frequency while that of the capacitor, X_C, decreases. The reactances are given by

$$X_L = 2\pi f L \qquad (1.6)$$

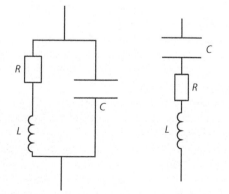

FIGURE 1.7 Parallel and series tuned circuits.

$$X_C = \frac{1}{2\pi f C} \tag{1.7}$$

At low frequencies, the capacitive reactance dominates the circuit by virtue of the inverse relationship with frequency. (The inverse of something small is something big.) As frequency increases, the capacitive reactance decreases, but the inductive reactance increases. There comes a frequency where they are both equal. This is called the *resonant frequency*, f_0, and it is a very important parameter:

$$X_L = X_C$$

Therefore,

$$2\pi f_0 L = \frac{1}{2\pi f_0 C}$$

And so

$$\omega_0 = \frac{1}{\sqrt{LC}} \tag{1.8}$$

At resonance the inductor will store energy while the capacitor will have none. Then, the situation reverses and the capacitor has the energy and the inductor has none. In the ideal case, the parasitic resistance, R, will be zero. So, the net current taken from the supply will be zero because the inductor has a phase shift of $+90°$ associated with it while the capacitor has a phase shift of $-90°$. Thus, they cancel each other out. If the current is zero, the impedance is infinity. This will not happen in practice because every inductor has a series resistance due to the windings that make it up, and so the impedance at resonance will not be infinite.

Figure 1.8 shows the current through a series and a parallel tuned circuit. Note that the scale used is arbitrary. For the series tuned circuit, the current is initially very low because the capacitor has a very large reactance and it is in series with the inductor. When resonance occurs, the total reactance is at its lowest value and so the current through the circuit is at a maximum. At a high frequency, the inductive reactance is high and so the current is low. For the parallel circuit, the current is at a maximum at low frequencies because it flows through the inductor. It reaches a minimum at resonance and then increases again as the high frequency causes the current to flow through the capacitor. The current at resonance is at a minimum because the inductor current equals the capacitor current, but there is a total phase shift between them of $180°$ and so the current is zero. At resonance, the parallel circuit has maximum resistance. It should be noted that the spread with frequency is due to resistance in the circuit and this is considered next.

A useful parameter when considering tuned circuits is Q, the quality or magnification factor of a circuit. This is 2π times the ratio of the total stored energy to the energy lost. Energy is lost in any resistance be it intentional or parasitic. There is a magnification of current in the tuned circuit due to the transfer of energy from the inductor to the capacitor and Figure 1.9 shows the current "amplification" in

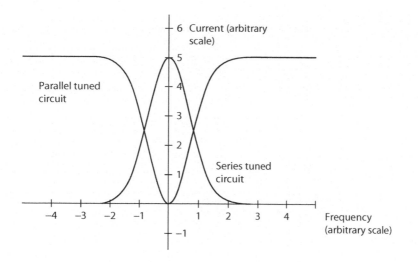

FIGURE 1.8 Current through a parallel and series tuned circuit.

a parallel tuned circuit. If the resistance of the inductor is zero, the amplification would be infinite and the range of frequencies passed would be zero. As the resistance increases, the current amplification reduces and the bandwidth goes up. The Q for a parallel circuit with an inductor with resistance is given by

$$Q = \frac{\omega_0 L}{R} \tag{1.9}$$

or

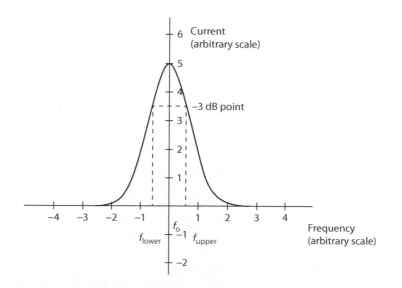

FIGURE 1.9 The bandwidth of a tuned circuit.

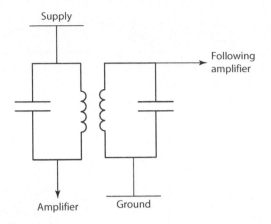

FIGURE 1.10   A coupled tuned circuit.

$$Q = \frac{f_0}{\Delta f} \qquad (1.10)$$

where Δf is the bandwidth – the difference between the upper and lower 3 dB points. From Equation 1.10 it is evident that Q of the circuit is a measure of how selective the circuit is. A high Q means that the v response of the circuit has a narrow width and this is extremely useful when selecting a particular radio station. Appendix III contains a mathematical derivation for these formulae.

Signals can be coupled from the tuned circuit by either taking it directly from the inductor/capacitor or by inductive coupling. It is the inductive coupling that is most effective because the inductor acts as the primary coil in a transformer configuration and the secondary coil can also be tuned (Figure 1.10).

Tuned circuits are generally housed in a metal can to shield other circuits from electromagnetic radiation from the inductor (coil). Figure 1.11 shows a typical tuned circuit both with and without the can. Note that there is a ferrite core in the coil which can be moved up and down inside it. This is to fine-tune the circuit. The dimensions are 1 cm × 1 cm × 1.3 cm.

1.7   BASIC RADIO RECEIVER DESIGN: THE SUPERHETERODYNE RECEIVER

As we will see later, the voltage at the terminals of an antenna can be very low (of the order of 1 mV) and we must amplify this voltage to a level that can be heard. This is the job of a series of amplifiers, mixers and a demodulator as shown in Figure 1.12.

The antenna picks up a range of frequencies, for instance the medium-wave band, and these are amplified by the two tuned radio frequency (RF) amplifiers. The resulting signals are then mixed to the intermediate frequency (IF) where they are filtered by the IF amplifiers. A demodulator converts the signal back into audio frequencies (AF) prior to amplification and then the loudspeaker. The operation is more easily explained by reference to Figure 1.13.

FIGURE 1.11   Tuned circuit in a metal can and the coil inside.

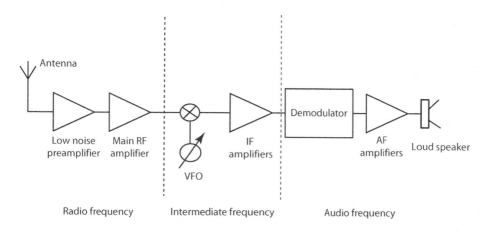

FIGURE 1.12   Block diagram of a superheterodyne receiver.

Figure 1.13a shows a number of radio stations as picked up by the receiver antenna. Note that there are a large number as the receiver antenna need not be too selective. The first two amplifiers in the receiver are RF amplifiers and they select a fairly narrow range of stations. Care must be exercised here so that the RF amplifiers do not filter out the station required. For this reason, the post-amplifier is often more selective than the pre-amplifier. In the diagram, only four stations are selected

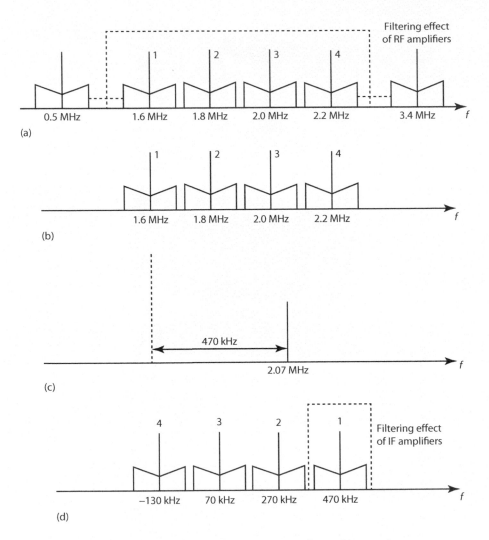

FIGURE 1.13 Filtering and frequency shifting in a superheterodyne receiver.

to pass to the mixer. The other input to the mixer is a local oscillator which can vary in frequency. This variable frequency oscillator (VFO) is tuned to a frequency of 1.6 MHz (Station 1) plus 470 kHz (Figure 1.13c). This offset is the IF (intermediate meaning intermediate between the RF and AF). The resultant mixing products are the sum and difference frequencies and Figure 1.13d shows the difference components. Tuning of the IF amplifiers, using parallel tuned circuits, to the IF of 470 kHz means that they will reject all stations that are not at the IF. Thus, Station 1 passes through to the demodulator and is amplified prior to conversion back into a pressure wave in the loudspeaker. Note that there is a negative frequency here – Station 4. This negative appears as a positive frequency of 130 kHz because the cosine of a negative number equals the cosine of a positive number. In effect, the negative signals fold around zero to appear positive. Care must be taken to ensure that the "negative" frequency does not interfere with the station sitting at 470 kHz, otherwise interference will result.

If we want to select Station 2 at 1.8 MHz, we need the VFO to be at $1.8 + 0.47 = 2.27$ MHz. The difference between the frequency of Station 2 and the VFO will result in Station 2 instead of Station 1 sitting at 470 kHz and so the receiver is tuned to Station 2.

1.8   THREE VERY IMPORTANT THEOREMS: NYQUIST (TWICE) AND SHANNON

Three very important theorems are introduced here rather than later, such is their importance. The first is the Nyquist rate, to do with how often we must sample a signal to avoid distortion. Sampling is nothing new to us. When we go to the cinema, we see moving pictures – individual frames are displayed so quickly that we do not notice the sampling. As we will see later (Chapter 3), this is of great interest in digital communications. The sampling rate, f_s, is governed by

$$f_s \geq 2 f_m \tag{1.11}$$

where f_m is the maximum modulating frequency. As an example, if we take an audio frequency of 3.4 kHz, the sampling frequency should be at least 6.8 kHz.

The second theorem relates to channel capacity. This is the Shannon–Hartley theorem which places a fundamental limit on the capacity of a transmission link, C, as

$$C = B \log_2 S/N \tag{1.12}$$

where:
   C is the channel capacity in bit per second
   B is the bandwidth of the channel
   S/N is the signal to noise ratio

Equation 1.12 shows that capacity can be increased if the S/N and the bandwidth are increased. (Although we are yet to meet some of these terms, the importance of these theorems means that they need introducing here.)

The third is due to Nyquist again and relates the bandwidth of a system to the data rate that can be accommodated:

$$B = 2 f_{\text{channel}} \tag{1.13}$$

where:
   B is the bit rate to be transmitted
   f_{channel} is the channel bandwidth in Hertz

So, ideally, a 100 MHz bandwidth channel could support 200 Mbit/s of data.

1.9 PROBLEMS

1. Determine the length of a half wavelength dipole for each of the following
   frequencies:
   a. 1 kHz
   b. 1 MHz
   c. 1 GHz
   d. 100 GHz
      [150 km; 150 m; 15 cm; 1.5 mm]
2. Convert the following into their sinusoidal form (i.e. $V \sin\omega t$):
   a. 2 V peak, 10 kHz
   b. 2 V peak-peak, 1 MHz
   c. 1 V rms, 50 MHz
   d. 5 V peak, 628.32 Mrad/s
      [$2 \sin 62.83 \times 10^3 t$; $1 \sin 6.28 \times 10^6 t$; $1.414 \sin 314.16 \times 10^6 t$;
      $5 \sin 628.32 \times 10^6 t$]
3. Confirm, by expanding $\sin (\omega t + \pi/2)$, that a cosine is simply a sine with a
   90° ($\pi/2$) phase shift.
4. Determine the values of the following sinusoids at their respective times:
   a. $10 \cos 2\pi 100 \ t$ at time $t = 1$ ms
   b. $100 \sin 2\pi 1 \times 10^5 \ t$ at time $t = 10$ μs
   c. $25 \sin (2\pi 1 \times 10^6 \ t + \pi/4)$ at time $t = 2$ μs
   d. $2 \cos (628.31 \ t + 3\pi/2)$ at time $t = 0.1$ ms
      [8.1 V; 0 V; 17.68 V; 0.126 V]
5. A parallel tuned circuit uses an inductance of value 10 mH with a parasitic
   resistance of 10 Ω. Calculate the resonant frequency and Q for the following
   capacitances:
   a. 10 nF
   b. 100 pF
   c. 1 pF
      [16 kHz, 100; 160 kHz, 1×10^3; 1.6 MHz, 10×10^3 – note that the values
   of Q are ideal. As the capacitance gets smaller, so the effect of parasitic
   capacitance increases and that makes it difficult to achieve the resonant
   frequency without reducing the inductance and hence the Q.]
6. A parallel tuned circuit is designed to have a resonant frequency of 10.7
   MHz. The value of the capacitance used is 100 pF. Determine the value of
   the inductor to be used and the Q factor if the parasitic resistance is 5 Ω.
      [2.2 μH, 30]

2 Noise

Noise places a limitation on telecommunications systems. It can be external to the receiver or internally generated. It is possible to reduce external noise by filtering and by making sure that the receiver and all sources of external interference are certified. (The process of certification measures the amplitude and frequency of noise produced by an artefact. It should mean that a radio can be used next to a vacuum cleaner without the cleaner affecting the radio with electrical noise.) Internally generated noise in a radio system comes from resistors and semiconductors. As amplifiers use these things, every radio system generates noise. It is possible to hear noise in hi-fi systems by turning up the volume without any input present of course!

There are two types of noise that we have to contend with – flicker noise ($1/f$ or pink noise) and white noise. White noise is so named because it is analogous to white light in that it has a constant amplitude with variation in frequency. It also has a spectrum that extends to infinity. White light has all the visible colours with equal amplitude. The amplitude of flicker noise is not constant with frequency, rather it has a $1/f$ distribution. It is also termed *pink noise* because light with a $1/f$ distribution looks pink. As flicker noise reduces with frequency and white noise is flat with frequency, there comes a point when the total noise is 3 dB above the white noise – the flicker noise corner frequency. This frequency can be as low as 2 kHz for a junction gate field-effect transistor (JFET) and bipolar transistors, but it can be as high as 1 GHz for *metal-oxide semiconductor field-effect transistors* (MOSFETs). It is normally due to changes in resistance and is associated with direct current (dc) voltages and currents. It is implicated in phase noise in oscillators and can cause frequency variations in voltage-controlled oscillators. We will ignore flicker noise in most of the work we do here, but it is as well to remember that it is there.

2.1 CIRCUIT NOISE: WHY AMPLIFIERS HISS

There are two types of noise in a radio receiver: thermal noise that is generated by resistors and shot noise that is generated by semiconductors.

Thermal noise comes from the thermal agitation of electrons in a resistor. This agitation is statistical in nature and so it is impossible to say for certain what the increase in voltage will be. This fluctuation in carriers causes noise with the thermal noise being expressed as a mean square noise voltage:

$$\langle v_n^2 \rangle = 4\mathrm{k}TRB \ \mathrm{V}^2 \tag{2.1}$$

$$= 4\mathrm{k}TR \ \mathrm{V}^2/\mathrm{Hz} \tag{2.2}$$

where:

    k   is Boltzmann's constant (1.38×10^{-23} J/K)

    T is the absolute temperature of the resistor in Kelvin

    R is the resistance in Ohm

    B is the bandwidth in Hertz

    $\langle \; \rangle$ indicates mean value

    subscript n indicates noise

Thus, $\langle v_n^2 \rangle$ is the mean square noise voltage. The units in Equation 2.1 are V^2 while those in Equation 2.2 are V^2/Hz. These latter units show that Equation 2.2 is a mean square noise voltage spectral density. This is a very useful representation as we will see later.

Shot noise is generated in semiconductors and it comes from the fact that carriers that conduct current in semiconductors are not evenly distributed in the conduction band. This randomness manifests itself as noise. (Shot noise is not just confined to semiconductors but also appears in thermionic valves.) The shot noise is expressed as a mean square noise current:

$$\langle i_n^2 \rangle = 2qIB \; A^2 \tag{2.3}$$

$$= 2qI \; A^2/Hz \tag{2.4}$$

where:

    q is the electronic charge (1.6×10^{-19} C)

    I is the current in amperes

    B is the bandwidth as before

Any meaningful theoretical analysis of an amplifier is hampered by the fact that there will be noise sources that are voltages (resistors) and currents (transistors) and these cannot be added together directly. It is possible to convert voltage to current (Ohm's law and circuit theorems) but it is more usual to simulate the amplifier using Software Program for In-Circuit Emulation (SPICE), a circuit simulation package of which there are many variations freely available.

The noise just described is termed *white noise* as it has a constant distribution with respect to frequency. The random amplitude has a Gaussian distribution and so the probability density function (pdf) is

$$p(v) = \frac{1}{\sqrt{2\pi\sigma^2}} \exp\left(\frac{-(v-a)^2}{2\sigma^2} \right) \tag{2.5}$$

where:

    $p(v)$ is the pdf of the noise voltage

    σ^2 is the mean square noise

    v is the noise voltage

    a is the average noise level, which is not always zero

Equation 2.5 will be used when we consider noise in baseband digital transmission in Section 3.2.

2.2 NOISE FACTOR AND FIGURE

The noise factor of a network (F) is defined as

$$F = \frac{S/N^{\ I/P}}{S/N^{\ OUT}} \tag{2.6}$$

where S/N is the signal to noise ratio at the input to, and output of, a network. (There are two alternative expressions for F but they are both related to Equation 2.6 and are not considered here.) The ratio S/N is a power ratio so it is the ratio of signal power to noise power. In a radio receiver, we are interested in the S/N performance because this will determine the fidelity of the demodulated signal. The noise figure (NF) is the noise factor expressed in decibels (Appendix IV).

$$NF = 10 \log_{10} F \ dB \tag{2.7}$$

There are essentially two components that make up a receiver: amplifiers and lossy components. The first is obvious because we need to amplify the signal to get it to a level that can be heard. Lossy components introduce attenuation so that the signal is diminished in power. This is not what we want to do; however, coaxial cable is widely used to connect the antenna to the receiver, and mixers can be passive (introducing a loss) as well as active (having a gain). The noise of an amplifier and lossy components can be measured using a noise figure meter.

Consider an amplifier with a gain of 10 dB and a noise figure of 3 dB. The gain as a ratio is 10 and the noise factor is 2.0. Any signal at the input to the amplifier is boosted by a factor of 10. Unfortunately, any noise on the input is also amplified by the same amount and so the S/N does not increase. In fact, the amplifier adds noise itself by virtue of shot and thermal noise. So, the S/N at the output of the amplifier is less than that at the input. This is where the noise factor of 2 comes in and the S/N at the output is reduced to half that at the input.

Now consider a length of coaxial cable with a loss of 6 dB (4 as a ratio). Any signal entering the coax will be attenuated by a factor of 4 and any associated input noise will also be attenuated by the same amount. The cable will introduce its own noise and so the S/N at the output will be less than at the input. As shown in Appendix V, the noise factor for a lossy network is equal to the loss. Thus, F would be 4 and so the S/N after the cable will be ¼ that at the input.

2.3 NOISE POWER FROM AN ANTENNA

Consider an antenna connected to a resistive load. This load could be a length of cable or an amplifier. The antenna can be modelled as a source, E_s, with an internal resistance, $R_{antenna}$, as shown in Figure 2.1. The load is modelled as a resistance R_{load}.

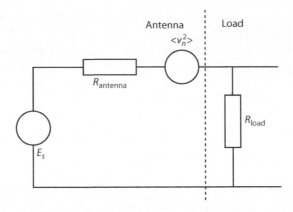

FIGURE 2.1   Equivalent circuit of an antenna connected to a load.

Maximum voltage transfer occurs with the load equal to infinity but this does not help receive the signal because the power is zero (no current). If the load is a short circuit, we have maximum current, zero volts and zero power again. Maximum power transfer occurs when the source and load resistances are the same (Appendix VI). So,

$$R_{\text{antenna}} = R_{\text{load}} = R \qquad (2.8)$$

The electromotive force (emf) will produce $E_s/2$ across the load, but remember that this is maximum power transfer. There are two resistors in a circuit but the noise produced by the load resistor is considered to be part of the load be it an amplifier or a length of cable. Thus, the only noise source is the antenna resistance, R. This will generate the thermal noise of Equation 2.2:

$$\langle v_n^2 \rangle = 4\text{k}TRB \text{ V}^2$$

This voltage will generate a circulating current, $\langle i_n^2 \rangle$, through the two resistors. So,

$$\langle i_n^2 \rangle = \frac{4\text{k}TRB}{(R+R)^2} \text{ A}^2$$

$$= \frac{4\text{k}TRB}{(2R)^2} \text{ A}^2$$

$$= \frac{4\text{k}TRB}{4R^2} \text{ A}^2$$

$$= \frac{\text{k}TB}{R} \text{ A}^2$$

This current flows through the load and so the noise power developed across the load is (I^2R):

$$P_{\text{noise}} = kTB \text{ W} \qquad (2.9)$$

It should be evident from Equation 2.9 that the noise power delivered by the antenna to the load is independent of resistance provided the source and load are matched. As an example, consider an antenna with a voltage of 1 mV root mean square (rms) across the terminals, a system bandwidth of 100 kHz, a temperature of 290 K and a system matched to 50 Ω. As the voltage is across the load, the signal power is

$$S = \frac{\left(1 \times 10^{-3}\right)^2}{50}$$

$$= 20 \text{ nW}$$

The noise power from the antenna is

$$N = 1.38 \times 10^{-23} \times 290 \times 100 \times 10^3$$

$$= 4 \times 10^{-16} \text{ W}$$

Thus, the S/N of the signal after it has been picked up by the antenna is

$$\frac{S}{N} = \frac{20 \times 10^{-9}}{4 \times 10^{-16}} = 5 \times 10^7 = 77 \text{ dB}$$

So, there is a finite S/N as soon as a signal is picked up by the antenna. Note that the mean square noise is directly proportional to the bandwidth and so large bandwidth signals suffer with a lot of noise. What happens to the S/N after the antenna is the subject of the next section.

2.4 CASCADED NETWORKS: FRISS' FORMULA

We have already seen that a radio receiver has many blocks associated with it – selective amplifiers, mixers, more amplifiers. This is known as a *cascaded network* and it is usually matched to typically 50 Ω for radio and 75 Ω for TV. The noise factor of a cascade, F_{total}, is given by Friss' formula (Appendix V):

$$F_{\text{total}} = 1 + \left(F_1 - 1\right) + \frac{\left(F_2 - 1\right)}{G_1} + \frac{\left(F_3 - 1\right)}{G_2 G_1} + \cdots \qquad (2.10)$$

where F_n and G_n are the noise figure and power gain of the nth stage (Figure 2.2) as a ratio not decibels.

Consider the receiver system shown in Figure 2.3. Let the received carrier generate a voltage of 1 mV rms across the antenna terminals and let the components have the noise figure and gain as shown in Figure 2.3. The decibel figures have to be converted to ratios:

$$dB = 10 \log\left(\text{power ratio}\right) \qquad (2.11)$$

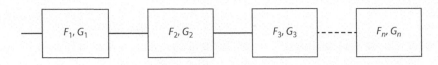

FIGURE 2.2   A cascaded network.

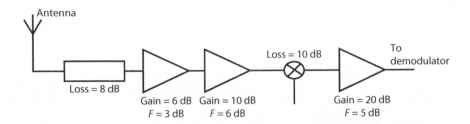

FIGURE 2.3   Block diagram of a receiver cascade.

TABLE 2.1
Gain, Noise Figure and Noise Factor of the Blocks in Figure 2.3

Component	Gain (dB)	Gain (Ratio)	Noise Figure (dB)	Noise Factor
Coaxial cable	−8	0.16	8	6.31
Amp 1	6	3.98	3	2.00
Amp 2	10	10.00	6	3.98
Mixer	−10	0.1	10	10
Amp 3	20	100.00	5	3.16

Hence,

$$\text{Power ratio} = 10^{dB/10} \qquad\qquad (2.12)$$

The easiest way to analyse the receiver is to draw up a table (Table 2.1).

The first stage is a length of coaxial cable which will attenuate the received signal. The amount of loss is 8 dB and so the gain is −8 dB. By using Equation 2.10, the total noise factor for the receiver is

$$F_{\text{total}} = 1 + (6.31 - 1) + \frac{2-1}{0.16} + \frac{3.98-1}{3.98 \times 0.16} + \frac{10-1}{10 \times 3.98 \times 0.16} + \frac{3.16-1}{0.1 \times 10 \times 3.98 \times 0.16}$$

$$= 1 + 5.31 + 6.31 + 4.71 + 1.41 + 3.39$$

$$= 22.13$$

In order to find the S/N at the input to the detector, we first need to find the S/N at the input to the receiver. This we have already done in the previous section, it is 5×10^7 or 77 dB. So, the S/N at the input to the detector is

$$\frac{S}{N} = \frac{S/N_{I/P}}{F} = \frac{5 \times 10^7}{22.13} = 2.26 \times 10^6 = 63.54 \text{ dB}$$

It should be noted that the cable, with its loss, has a major impact on the noise factor; it introduces noise of its own and it effectively amplifies the noise of the following stages – the noise is divided by 0.16 and so it gets bigger. A similar thing happens with the mixer (passive with a gain of 0.1). It should be obvious that a large preceding gain will reduce the noise when it is referred back to the input.

2.5   NOISE TEMPERATURE AND DIRECTIONAL ANTENNAE

Noise temperature is a convenient way of examining the noise performance of a network. In essence, it is the temperature at which a resistor equal to the system resistance (50 Ω for example) must be, so that it produces the same amount of noise as the component it replaces. From Equation AV.5 in Appendix V, we know that the amplifier noise can be given by

$$N_a = (F - 1)GkTB \tag{2.13}$$

The effective equivalent noise temperature is

$$T_e = (F - 1)T \tag{2.14}$$

The temperature, T, in Equation 2.14 is the physical temperature of the component and is usually taken to be 290 K. When it comes to the temperature of the antenna, it is not the physical temperature that we work with. Consider a long wire antenna with no directivity. In this case, the antenna receives radiation from all directions. Thus, its noise performance is that of a resistor at the accepted earth temperature of 290 K. If the antenna is directional such as those found in high-frequency links, the effective noise temperature is a lot lower and could be as low as 50 K. This is due to it only "seeing" a section of sky, hence its noise is lower.

It is possible to use the method in Section 2.4 to find the S/N. However, the noise temperature approach can give more insight into the build-up of noise in a receiver. Taking the previous example but with an antenna temperature of 50 K gives Table 2.2.

It is a simple matter to show that the overall effective input noise temperature is

$$T_{etotal} = T_{eant} + T_{e1} + \frac{T_{e2}}{G_1} + \frac{T_{e3}}{G_2 G_1} + \cdots \tag{2.15}$$

TABLE 2.2

Noise Temperatures for the Components in Figure 2.3

Component	Gain Decibel to Ratio	Noise Decibel to Ratio	Te (K)
Antenna	–	–	$T_{eant}=50$
Coaxial cable	$-8=0.16$	$8=6.31$	$T_{e1}=1540$
Amp 1	$6=3.98$	$3=2.00$	$T_{e2}=290$
Amp 2	$10=10.00$	$6=3.98$	$T_{e3}=870$
Mixer	$-10=0.1$	$10=10$	$T_{e4}=2610$
Amp 3	$20=100$	$5=3.16$	$T_{e5}=626$

Thus

$$T_{etotal} = 50+1540+\frac{290}{0.16}+\frac{870}{3.98\times0.16}+\frac{2610}{10\times3.98\times0.16}+\frac{626}{0.1\times10\times3.98\times0.16}$$

$$= 50+1540+1812+1366+410+983$$

$$= 6161\,\text{K}$$

Note how the temperatures vary. The antenna temperature is very low but the coaxial cable has a very high temperature. The first stage amplifier is actually the major source of noise even though its temperature by itself is quite low. This is due to the noise being referred through the lossy coaxial cable. The solution is to use an amplifier right at the antenna terminals – a mast-head pre-amplifier – and this helps to mask the effect of the cable.

To calculate the S/N, we note that the effective input noise temperature of the receiver is the temperature that a resistor (50 Ω) must be at to generate the same amount of noise as the receiver. So, the receiver will generate noise of

$$kT_{etotal}B = 1.38\times10^{-23}\times6161\times100\times10^{3}$$

$$= 8.5\times10^{-15}\ \text{W}$$

From the previous section, Section 2.3, the signal is 20 nW and so the S/N is

$$\frac{S}{N} = \frac{20\times10^{-9}}{8.5\times10^{-15}} = 2.35\times10^{6}$$

Comparing to the previous example (S/N=2.26×10^{6}) reveals that there is little change in the S/N. This is because we only changed the antenna and there is not much to be gained if we don't alter the receiver.

We are now able to perform system calculations on receivers up to the demodulator. In order to examine the noise performance of amplitude modulation (AM) and frequency modulation (FM) detectors, we must first discuss the algebraic representation of noise.

2.6   ALGEBRAIC REPRESENTATION OF NOISE: FILTERED NOISE

In carrier-based systems, the noise is filtered as it passes through the receiver and, in particular, the intermediate frequency (IF) stages. Figure 2.4 shows a representation of the passband of the IF stage. Recall that it is this stage that provides the selectivity so that a particular radio station is received. The response of the IF stage has been approximated to that of an ideal bandpass filter centred on f_{IF} and with a bandwidth B.

Consider noise from the receiver being filtered by the IF stage and let the noise spectral density be η W/Hz ($\eta = kT$). The noise in a strip Δf wide centred on a frequency f is $\eta \Delta f$. As Δf tends to df, the component at frequency f tends to a straight line which represents a single frequency, i.e. $V\cos\omega t$. The power in this sinusoid must equal that of the power in the strip. So,

$$\eta df = \left(\frac{V}{\sqrt{2}} \right)^2$$

On rearranging, this gives $V = \sqrt{2\eta df}$.

Thus, the noise at frequency f can be written as

$$v_n(t) = \sqrt{2\eta df}\,\cos\omega t \tag{2.16}$$

Signals at the IF are translated as zero Hertz whereas those either side translate to a baseband frequency. For instance, at an IF of 470 kHz, a frequency of 470 kHz equates to 0 Hz on demodulation and a frequency of 471 kHz (or 469 kHz) equates to 1 kHz. So, what is important is not the absolute frequency but the offset from the IF. We can introduce an offset frequency, ω_{off}, given by

$$\omega_{off} = \omega - \omega_{IF} \tag{2.17}$$

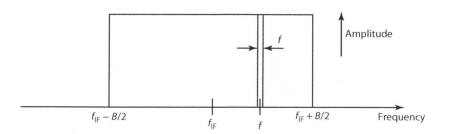

FIGURE 2.4   The IF response of a receiver as used for noise calculations.

Substituting this into Equation 2.16 yields

$$v_n(t) = \sqrt{2\eta df}\, \cos(\omega_{\text{off}} + \omega_{\text{IF}})t$$

$$= \sqrt{2\eta df}\, \{\cos\omega_{\text{off}}t . \cos\omega_{\text{IF}}t - \sin\omega_{\text{off}}t . \sin\omega_{\text{IF}}t\}$$

$$= \sqrt{2\eta df}\, \{\cos\omega_{\text{off}}t . \cos\omega_{\text{IF}}t\} - \sqrt{2\eta df}\, \{\sin\omega_{\text{off}}t . \sin\omega_{\text{IF}}t\} \qquad (2.18)$$

$$= x(t) . \cos\omega_{\text{IF}}t + y(t) . \sin\omega_{\text{IF}}t$$

Examination of Equation 2.18 shows that there are two terms to the noise both with the same amplitude. What should also be obvious is that the two terms are in quadrature – one is cosine and the other is sine. These are known as *in-phase* and *quadrature components* and it depends on how we draw the phasor diagram as to which is in-phase and which is quadrature. This will be useful when we consider carrier modulation in Section 3.3 and Chapter 4.

2.7   PROBLEMS

1. Convert the following power ratios into decibels: 0.1; 2; 10; 20; 100; 10^4.
   [−10; 3; 10; 13; 20; 40]
2. Convert the following decibels into power ratios: −8; 2; 5; 10; 15; 25; 40.
   [0.16; 1.58; 3.16; 10; 31.6; 316; 10^4]
3. A 100 MHz signal develops a voltage of 2 mV rms across the loaded terminals of a receiving antenna. If the system is matched to 50 Ω, the antenna temperature is 290 K and the system bandwidth is 100 kHz, determine the S/N.
   [1×10^8 or 80 dB]
4. The noise figure of a 50 Ω receiver, from antenna terminal to demodulator input, is 12 dB. Determine the carrier to noise (C/N) at the input to the demodulator if the C/N at the input to the receiver is 72 dB.
   [60 dB or 1×10^6]
5. In Table 2.1, the coaxial cable develops a fault which causes the loss to increase to 20 dB. Determine the new noise temperature and the new C/N at the input to the demodulator.
   [87600 K; 1.7×10^5 or 52.2 dB]
6. Noise power can be referred to as one Watt (dBW) or one milliwatt (dBm). Convert the noise power kTB to a noise spectral density (W/Hz) and then convert this to dBW and dBm. (Take a temperature of 290 K.) By using logs, determine the noise in a 100 kHz bandwidth.
   [4×10^{-21} W/Hz; −204 dBW; −174 dBm; −154 dBW]

3 Introduction to Digital Modulation

3.1 PULSE CODE MODULATION: DIGITISING SIGNALS

Telephony, landline and mobile phones, and digital TV are digital in form. The analogue signals are sampled and then digitised prior to transmission as binary digits (bits). The digitisation occurs within an analogue to digital converter (ADC) and the resulting data stream is commonly referred to as pulse code modulation (PCM). The process of sampling is well known to us – a TV picture is not continuous and the films we see at the cinema are composed of discrete frames. The same can apply to sound as our ears are unable to distinguish between sampled and continuous audio – if the samples are close enough, we will not be able to tell the difference. The Nyquist sampling theorem (Section 1.8) says that we need at least two samples per sine wave in order to recover the signal, i.e.

$$f_s \geq 2f_m \tag{3.1}$$

where:
  f_s is the frequency at which samples are taken
  f_m is the maximum frequency to be sampled

Speech is limited to 3.4 kHz in the telephone network and so the sampling frequency could be 6.8 kHz. However, the equality in Equation 3.1 assumes ideal low-pass filters. To operate with real parameters, the sampling frequency used in practice is 8 kHz.

Following on from the sampler is the ADC which converts the analogue samples into a digital representation. This is shown in Figure 3.1, where eight samples per cycle and four bits are used for coding. As can be seen from Figure 3.1, each sample generates a binary code. Working from the first sample, the codes are 1001, 1101, 1111, 1100, etc. When the signal goes negative, the digital word begins with a 0. A very important point to note about the coding of Figure 3.1 is that there are rounding errors. The second sample falls above the coding level 1101 but it is coded as 1101. A similar thing occurs with sample 4 – it is rounded down to the level corresponding to 1100. A different rounding error occurs with the eighth sample. This sample lies between two levels and so it can be either rounded up to 0011 (decimal 3) or rounded down to 0010 (decimal 2). This rounding error appears as noise because when the level is reconstructed in the decoding digital to analogue converter (DAC) the level will be set but it could have been higher or lower by half a level. This is known as *quantisation noise* because of the similarity to quantum levels in physics. (In the model of the atom, only certain energy levels are allowed and so energy levels are quantised.) Figure 3.2 shows the rounding error in greater detail.

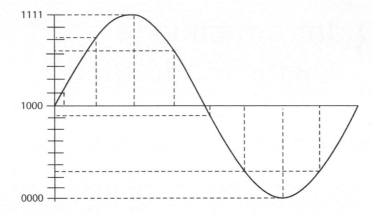

FIGURE 3.1   Sampling of a sine wave.

Figure 3.2 shows three quantisation levels at 100 mV (0001), 200 mV (0010) and 300 mV (0011). Any signal that lies between the first two levels, between 150 and 200 mV, is rounded up to 200 mV (0010). If the signal lies between 200 and 250 mV, the signal is rounded down to 200 mV (0010). Thus, the code 0010 could have been generated by 200 ± 50 mV. This error voltage is random in nature and so appears as noise – quantisation noise. The amount of quantisation noise can be found as follows:

For every voltage level coming out of the DAC there is an uncertainty associated with it of ±$\Delta V/2$. So, the error voltage, ε, is

$$-\Delta V / 2 < \varepsilon < +\Delta V / 2 \tag{3.2}$$

where ΔV is the step size (100 mV in the previous example). ΔV is given by

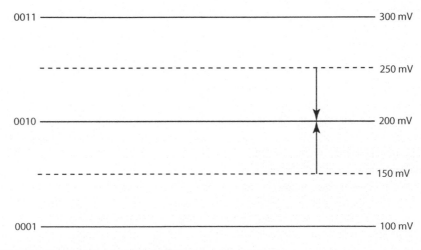

FIGURE 3.2   Quantisation in an ADC.

$$\Delta V = \frac{2V}{2^n - 1} \tag{3.3}$$

where:

2V is the range of the ADC

n is the number of bits used in the ADC

The mean square (ms) quantisation noise, N_q^2, is

$$N_q^2 = \frac{1}{\Delta V} \int_{-\Delta V/2}^{\Delta V/2} \varepsilon^2 d\varepsilon$$

$$= \frac{1}{\Delta V} \left(\frac{\Delta V^3}{8} + \frac{\Delta V^3}{8} \right) \frac{1}{3}$$

$$= \frac{1}{\Delta V} \frac{\Delta V^3}{12}$$

Therefore,

$$N_q = \frac{\Delta V}{\sqrt{12}} \tag{3.4}$$

To find the maximum signal to noise (S/N) ratio, we note that the range of the ADC is from $-V$ to $+V$. So, the maximum possible root mean square (rms) signal is simply

$$S = \frac{V}{\sqrt{2}} \tag{3.5}$$

So,

$$\frac{S}{N_q} = \frac{\frac{V}{\sqrt{2}}}{\frac{\Delta V}{\sqrt{12}}}$$

$$= \frac{V}{\sqrt{2}} \frac{\sqrt{12}}{\Delta V} \tag{3.6}$$

$$= \frac{V}{\sqrt{2}} \frac{\sqrt{12}}{2V} (2^n - 1)$$

$$= \frac{\sqrt{3}}{\sqrt{2}} (2^n - 1)$$

It is evident from Equation 3.6 that the more bits used to code the data, the greater the S/N_q. This is an obvious conclusion because the quantisation noise depends on the step size which in turn depends on the number of bits. As an example, consider an 8-bit ADC coding an input signal equal to the range of the ADC. The S/N_q is 49.9 dB (here the decibels are 20log ratio as we are dealing with voltages).

If the signal voltage is reduced, the S/N_q will fall. A halving of the signal voltage will reduce the S/N_q by 6 dB and further reductions will mean that there will come a point at which the quantisation noise is so great as to cause a problem to the listener. For speech (as in a telephone), this comes at an S/N_q of 30 dB. Thus, the dynamic range in this example is almost 20 dB. Speech is generally accepted to have a dynamic range of 40 dB, so we have to decrease the quantisation noise somehow. An obvious solution would be to increase the number of bits, so reducing the step size and the noise. However, this will increase the bit rate down the line. The solution used in practice is to introduce a non-linear amplifier before the ADC to compress the audio and then use a further non-linear amplifier to expand the signal – called *compression* and *expanding* or *companding*. There are two standards in use – μ law in America, Japan and Canada and A law which is used in Europe and the rest of the world. Internationally, if one country uses A law companding, the call will be made with A law. If two countries use the same μ law then μ law is used.

The μ law compression rule is

$$y = \frac{\ln(1+\mu x)}{\ln(1+\mu)} \quad 0 \le x \le 1 \tag{3.7}$$

where $\mu = 255$.

The A law rule is

$$y = \frac{1+\ln Ax}{1+\ln A} \quad 1/A \le x \le 1$$

$$y = \frac{Ax}{1+\ln A} \quad 0 \le x \le 1/A \tag{3.8}$$

with $A = 87.6$.

Non-linear quantisation basically uses a large number of linearly spaced quantum levels around zero, where small amplitude signals are, and then spacing the remaining levels on a logarithmic basis. Thus, quiet speakers have lots of levels, and so small quantisation noise, while the loud signals have fewer levels spaced further apart giving larger quantisation noise on a large signal. In this way, the S/N_q can be made reasonably constant over the range of the ADC.

Figure 3.3 shows the μ law and A law response and, as can be seen, they are almost identical. This does not mean that they are interchangeable because the coding of the PCM data following the 8-bit ADC is different. By using companding, however, the dynamic range of the ADC can be matched to that of speech. The data rate of a single PCM channel is simply given by the product of the number of bits and the sampling frequency, which is 8 bits times 8 kHz, giving 64 kbit/s.

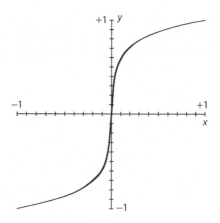

FIGURE 3.3   μ-law and A-law companding.

3.2   BASEBAND DIGITAL SIGNALLING: DATA TRANSMISSION

The 64 kbit/s data stream coming from our single speech channel is time-division multiplexed (TDM) onto a faster data stream that could contain further audio channels, video channels, Internet data, etc. In fact, we do not care what the data represents, we are only concerned with getting the signals to their destination. Let us consider a simple digital link such as point-to-point transmission.

Consider a binary digital link with signal amplitudes of 0 and V for logic 0 and 1. Noise will add to the signal and so the job of the detector is to decide whether a logic 0 or a logic 1 was transmitted. As shown in Figure 3.4, there are three basic blocks to a digital detector: a threshold crossing detector (comparator); a decision gate (D-type flip-flop); and a timing extraction circuit (phase-lock loop [PLL]). The comparator slices the data so that the data presented to the flip-flop is at the correct voltage level.

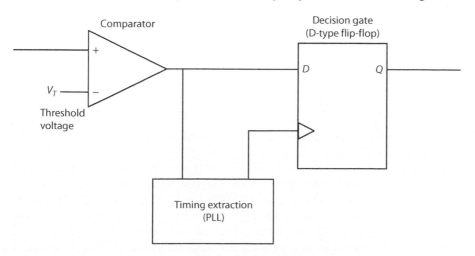

FIGURE 3.4   Simplified block diagram of a digital receiver/regenerator.

The D-type flip-flop retimes the data using the clock produced by the PLL. As there is attenuation along a line, the signal will need amplifying and filtering before the comparator. This is covered later in this section when we consider noise. One thing worthy of note is that the components operate at the baud rate, which may or may not be the same as the data rate.

Of interest in data communications is the error rate and the bandwidth. The bandwidth is very important as data can be carried over relatively low bandwidth cables. Baseband data generally comes in two signalling formats: non-return-to-zero (NRZ) and return-to-zero (RZ), as shown in Figure 3.5. Note that there are two times of importance here: the bit time, T_b, and the pulse time, T_p. For the RZ pulses, the pulse time is generally half that of the bit time.

The spectrum of a pulse, $p(t)$, is obtained by performing the Fourier transform (FT). Thus,

$$p(t) = V \quad \text{for } -\frac{T}{2} \le t \le \frac{T}{2}, \ p(t) = 0 \text{ otherwise}$$

$$\text{FT of } p(t) = P(f) = \int_{-\infty}^{\infty} p(t) \exp(-j\omega t) dt$$

$$= \int_{-T/2}^{T/2} V \exp(-j\omega t) dt$$

$$= \frac{V}{-j\omega} \left| \exp\left(-\frac{j\omega T}{2}\right) - \exp\left(\frac{j\omega T}{2}\right) \right|$$

$$= \frac{V}{-j\omega} \left| \cos\frac{\omega T}{2} - j\sin\frac{\omega T}{2} - \cos\frac{\omega T}{2} - j\sin\frac{\omega T}{2} \right| \tag{3.9}$$

$$= \frac{V}{\omega} \left| 2\sin\frac{\omega T}{2} \right|$$

$$= VT \frac{\sin\frac{\omega T}{2}}{\omega T / 2}$$

So, the spectrum of the pulse is a sin x/x response and this is shown in Figure 3.6 for the NRZ and RZ pulses.

It should be obvious from Figure 3.6 that the spectrum of a pulse passes through zero at multiples of $1/T$. Also, the tails and precursors carry on to infinity on both sides. However, it is possible to limit the bandwidth with some loss of information. (In actual fact, we lose the sharp edges on the pulse so that it has finite rise and fall times. The more the bandwidth is limited, the slower the rise and fall times.) Also evident from Figure 3.6 is that the bandwidth requirement for RZ is double that of NRZ. Thus, NRZ is preferred when operating with high-speed data. The null at $1/T$

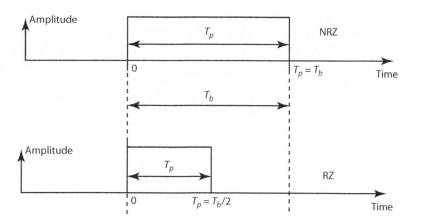

FIGURE 3.5   Timing diagram for NRZ and RZ pulses.

means that it is difficult to extract a clock at the frequency of the data. (Clock extraction is needed for retiming the data in the D-type flip-flop.) Non-linearities help here as the first null can be moved to a higher frequency thereby leaving a frequency component at the clock frequency which can be extracted.

Now let us examine how noise affects the detection of binary signals. One measure of performance is the error rate, measured in error bits per number of bits transmitted. Another is the error probability which is the probability of receiving an error.

We consider binary signalling with levels 0 and V volts. The additive noise acts to change a zero into a one and a one into a zero. We are dealing with white noise here because we do not have mixers and intermediate frequency (IF) amplifiers to

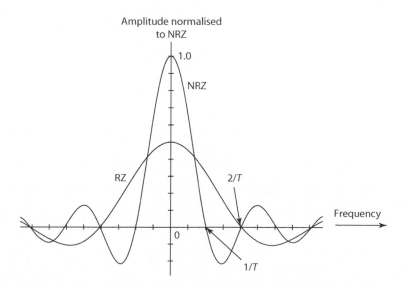

FIGURE 3.6   Spectrum of NRZ and RZ pulses.

band-limit the noise. So, taking Gaussian noise with a probability density function (pdf), as defined by Equation 2.5, on 0 and V gives the pdf of Figure 3.7.

As can be seen from Figure 3.7, the threshold voltage, V_T, for the comparator is set midway between logic 0 and logic 1. This assumes that there is an equal probability of a 1 or a 0 – they are equiprobable. This is a reasonable assumption because most transmission links are designed to have equiprobable symbols to avoid effects such as baseline wander. The next point to note is that every voltage above V_T is registered as logic 1 and everything below V_T is registered as logic 0. The Gaussian distributions on 0 and V represent the noise in the system. The two areas of interest are to the left and right of V_T. The area to the right represents a logic 1 being generated by noise on the logic 0 causing a false threshold crossing. Similarly, the area to the left of V_T is noise corrupting a pulse (logic 1) so that the threshold crossing is not registered. The crossover region is of interest because that is where the detection errors occur.

If the noise on a logic 1 is the same as the noise on a logic 0, the two areas of interest are the same (assuming V_T is halfway between the two logic levels). So, the total error probability, P_e, is

$$P_e = P_0 P_{e0-1} + P_1 P_{e1-0} \tag{3.10}$$

where P_0 and P_1 are the probabilities of a zero and one, respectively, and P_{e0-1} and P_{e1-0} are the probabilities of a zero going to a one and a one going to a zero, respectively. Taking the probability of a zero to be the same as that of a one, and assuming that the noise distributions are the same, P_e becomes

$$P_e = \frac{1}{2} P_{e0-1} + \frac{1}{2} P_{e0-1}$$
$$= P_{e0-1} \tag{3.11}$$

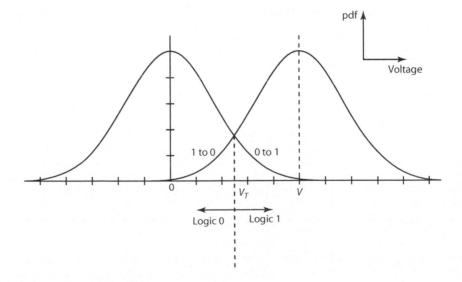

FIGURE 3.7 Gaussian noise on logic levels 0 and 1.

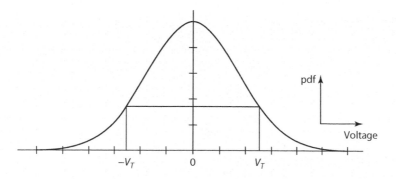

FIGURE 3.8 Construction to find P_e.

This is the tail under the logic 1 curve due to noise on the logic 0. To find P_e, we proceed as follows:

With reference to Figure 3.8, we require the tail from V_T to infinity. This can be found from

$$2P_e = \int_{-\infty}^{+\infty} p(v)\,dv - \int_{-V_T}^{+V_T} p(v)\,dv$$

and so

$$2P_e = 1 - 2\int_0^{+V_T} p(v)\,dv$$

Thus,

$$P_e = \frac{1}{2}\left(1 - 2\int_0^{+V_T} p(v)\,dv\right) \tag{3.12}$$

The noise is Gaussian in form (Equation 2.5) with an average of zero. So,

$$p(v) = \frac{1}{\sqrt{2\pi}\sigma}\exp\left(-v^2/2\sigma^2\right)$$

Substituting this into Equation 3.12 yields

$$P_e = \frac{1}{2}\left(1 - 2\int_0^{+V_T} \frac{1}{\sqrt{2\pi}\sigma}\exp\left(-v^2/2\sigma^2\right)dv\right) \tag{3.13}$$

A substitution can be made in the integral in Equation 3.13. Thus,

$$P_e = \frac{1}{2}\left(1 - 2\int_0^{+V_T} \frac{1}{\sqrt{2\pi}\sigma}\exp\left(-v^2/2\sigma^2\right)dv\right)$$

$$u^2 = v^2/2\sigma^2 \quad u = v/\sqrt{2}\sigma \quad du = dv/\sqrt{2}\sigma$$

and so

$$P_e = \frac{1}{2}\left(1 - 2\int_0^{+V_T/\sqrt{2}\sigma} \frac{1}{\sqrt{2\pi}\sigma}\exp\left(-u^2\right)\sqrt{2}\sigma\, du\right)$$

$$= \frac{1}{2}\left(1 - \int_0^{+V_T/\sqrt{2}\sigma} \frac{2}{\sqrt{\pi}}\exp\left(-u^2\right)du\right) \tag{3.14}$$

$$= \frac{1}{2}\left(1 - \mathrm{erf}\left(V_T/\sqrt{2}\sigma\right)\right)$$

In Equation 3.14, the term erf is the error function, which is widely tabulated (Appendix VII). The term σ is simply the rms noise in the receiver. The argument is a signal to noise term and if the argument increases, erf tends to 1 and the error rate becomes very small. Thus, to achieve a small error rate we need a high peak-to-peak voltage, leading to a high threshold voltage (half the peak-to-peak voltage) and/or low noise. This is intuitively obvious.

There is still a question concerning the pulse shape, inter-symbol interference (ISI) and noise. This is covered by studying the pre-detection filter that is placed just before the threshold detector. Ideally, we would like the filter to have a flat response from direct current (dc) to half the bit rate. This would give sin x/x pulses and result in the lowest amount of noise and ISI. However, the ISI is only zero at the sampling point and that requires an extremely accurate clock. (In practice, there is always some uncertainty over the precise rising edge of the clock. This is known as jitter.) We cannot use an ideal filter because of the ISI issue and it cannot be realised. The solution is a compromise in the form of a filter with a raised-cosine transfer function. Such a filter is described by

$$H(f) = \frac{1}{2}\left[1 + \cos\left(\frac{\pi f}{B}\right)\right] \quad \text{with } f < B \tag{3.15}$$

with $H(f) = 0$ otherwise. This is known as a *full raised-cosine filter* and the transfer function is shown in Figure 3.9 together with that of an ideal low-pass filter. The pulse shape for the impulse of such a filter is

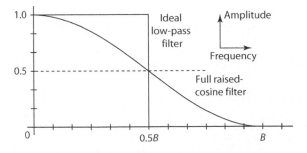

FIGURE 3.9   The frequency response of an ideal low-pass filter and a raised-cosine filter.

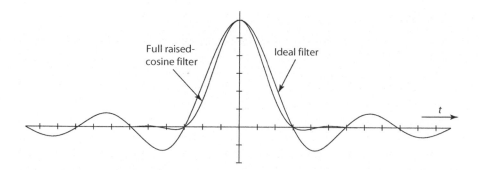

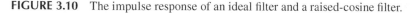

FIGURE 3.10 The impulse response of an ideal filter and a raised-cosine filter.

$$h(t) = \left(\frac{\sin \pi Bt}{\pi Bt} \right) \frac{\cos \pi Bt}{1 - (2Bt)^2} \qquad (3.16)$$

which is shown in Figure 3.10 together with the ideal sin x/x response. Note from Equation 3.16 that the sin x/x response has been modified by the cosine term and that causes the pre- and post-cursors to be significantly reduced, thereby alleviating the problem of ISI but at the expense of increasing the noise because of a higher bandwidth.

A major disadvantage of using baseband transmission for broadcast purposes is that only one transmission can be used. There is also the question of taking up a large part of the spectrum with just one station as well as different modes of propagation as frequency increases. So, a data transmission could potentially have half of its spectrum going right round the world while the other half is local to the transmitter. A solution that should be familiar to anyone using a networked computer is to use cable as a transmission medium or to use a carrier-based system, which is discussed next.

3.3 CARRIER-BASED SIGNALLING

Baseband signalling cannot be used for radio broadcasting because the bandwidth of the signals is simply too great and only one transmission is possible. Instead, a carrier-based approach can be used and there are several techniques that can be used to increase the data rate. Consider the carrier, $cv_c(t)$, represented by

$$v_c(t) = V_c \cos (2\pi f_c t + \varphi_c) \qquad (3.17)$$

There are three quantities that can be varied: the amplitude, V_c, giving amplitude modulation (AM); the frequency, f_c, giving frequency modulation (FM); and the phase, φ_c, giving phase modulation (PM). All of these can be used for digital communications – amplitude shift keying (ASK), frequency shift keying (FSK) and phase shift keying (PSK). Each "station" can use a different carrier frequency in the same way that different radio stations use different frequencies.

3.3.1 ASK

ASK is obtained by simply switching the carrier in and out, as in Figure 3.11. If the data is in the NRZ format, the spectrum is as in Figure 3.11. This is the NRZ spectrum except that it is at the carrier frequency – the carrier has been modulated by the data.

Demodulation (Figure 3.12) is simply a matter of mixing the ASK down to the baseband. This mixing produces sum and difference frequencies as noted in Appendix II. The sum component is at a high enough frequency to be easily filtered out. The difference component puts sidebands either side of zero, but it should be remembered that the cosine of a negative number is the same as the cosine of a positive number and so the signal folds around 0 Hz and reinforces the baseband.

As regards the noise performance, this is a carrier-based system and so we need to use the algebraic representation of noise in that there are in-phase and quadrature components. Thus, the received signal prior to the mixer is

$$v_{ASK}(t) = V_c \cos\omega_c t + x(t)\cos\omega_c t + y(t)\sin\omega_c t \ \text{ for } [1]$$

$$= 0 + x(t)\cos\omega_c t + y(t)\sin\omega_c t \ \text{ for } [0]$$

(3.18)

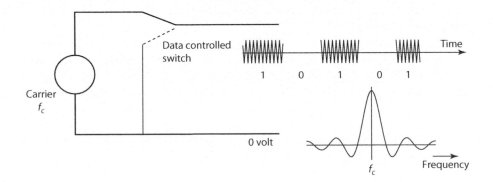

FIGURE 3.11   Generation of amplitude shift keying, the time plot and the spectrum.

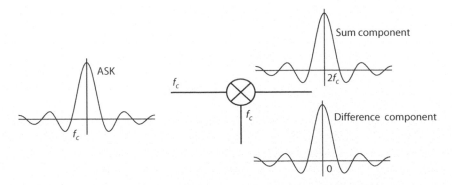

FIGURE 3.12   Demodulation of ASK.

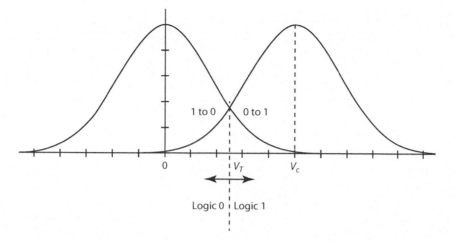

FIGURE 3.13 Gaussian noise on ASK levels 0 and 1.

This signal is translated to the baseband by mixing with $V_{LO}\cos \omega_c t$. This gives components at the baseband and $2\omega_c$. It is easy to check that the quadrature component generates zero and $2\omega_c$ terms and so will have no effect. Thus, after filtering, the ASK signal will be (Figure 3.13)

$$\text{ASK}(t) = (V_c + x(t))\cos \omega_c t . V_{LO} \cos \omega_c t$$

$$= \frac{V_{LO}}{2}(V_c + x(t)) \quad \text{for [1]} \quad \text{and} \quad \frac{V_{LO}}{2} x(t) \quad \text{for [0]}$$

We can ignore the $V_{LO}/2$ amplitude term as it multiples the amplitude and noise. Thus,

$$\text{ASK}(t) = V_c + x(t) \text{ for } \big[1\big] \text{ and } x(t) \text{ for } \big[0\big] \tag{3.19}$$

This is an identical situation to that encountered with baseband signalling presented in Section 3.2. So, by following a similar procedure, the error probability is

$$P_e = \frac{1}{2}\left(1 - \text{erf}\left(\frac{V_c}{2\sqrt{2\sigma^2}}\right)\right) \tag{3.20}$$

with σ^2 the ms noise voltage. As an example, consider V_c to be 500 mV and σ to be 50 mV. The argument for erf is 3.54 and so the probability of error is 273×10^{-9}.

3.3.2 FSK

In FSK, the modulating data switches between two different carriers, f_1 and f_2. As can be seen from Figure 3.14, the spectrum resembles two data streams at two different frequencies. Thus, the bandwidth is higher than ASK. Demodulation requires two demodulation paths both of which demodulate their respective signals to the

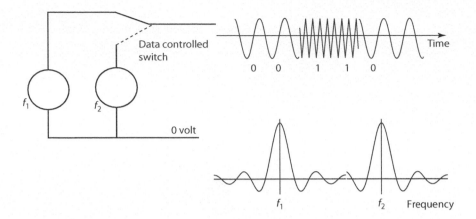

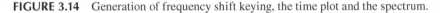

FIGURE 3.14 Generation of frequency shift keying, the time plot and the spectrum.

baseband (Figure 3.15). Consider the f_1 arm of the demodulator. The input signal is $V_c \cos \omega_1 t$ or $V_c \cos \omega_2 t$. Mixing with ω_1 results in sum and difference components which, after filtering, give a baseband for [0] and a similar argument applies for the other arm. The summer generates $+V$ [1] and $-V$ [0]. Noise adds to these levels giving the pdf of Figure 3.16 and, as can be seen, the threshold level is now at 0 V. The difference between the threshold and the logic levels is V_c, which should be compared to $V_c/2$ for ASK. The signal has been doubled and so the error rate should be a lot less. In principle this is true; however, there are two receivers here and they are summed at the end. Thus, the ms noise is doubled. (When one path has a pulse present plus noise, the other path has no signal but does have noise.) So, by following the same procedure as for the ASK receiver, the probability of error for FSK is

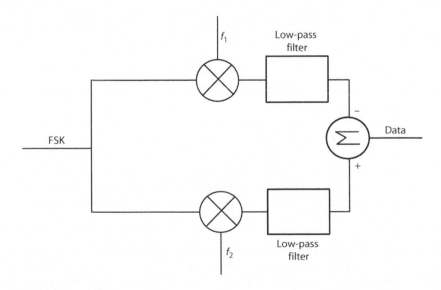

FIGURE 3.15 An FSK demodulator.

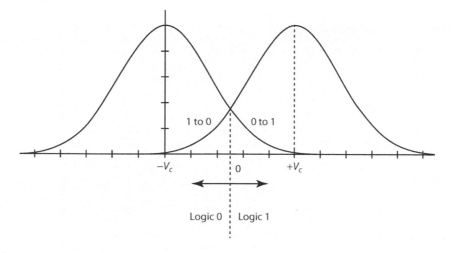

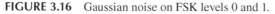

FIGURE 3.16 Gaussian noise on FSK levels 0 and 1.

$$P_e = \frac{1}{2}\operatorname{erfc}\left(V_c \middle/ \sqrt{2\times 2\sigma^2}\right)$$

(3.21)

Comparing with the previous example, the carrier voltage can be reduced to 354 mV for the same amount of noise and the same probability of error.

3.3.3 BPSK

With binary phase shift keying (BPSK), the phase of the signal varies according to the data. A logic one causes the phase to shift by 180°. As shown in Figure 3.17, the spectrum of BPSK is the same as for ASK and that makes it a very popular form of modulation. Demodulation is a case of mixing the BPSK signal down to the baseband; however, the carrier in the demodulator must be stable both in terms of frequency and, very importantly, phase. The Costas loop (Figure 3.18) can be used to demodulate the data.

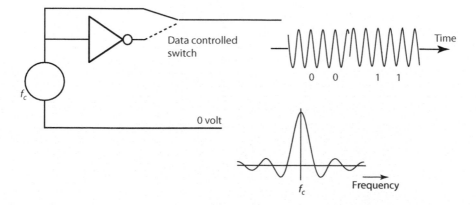

FIGURE 3.17 Generation of binary phase shift keying, the time plot and the spectrum.

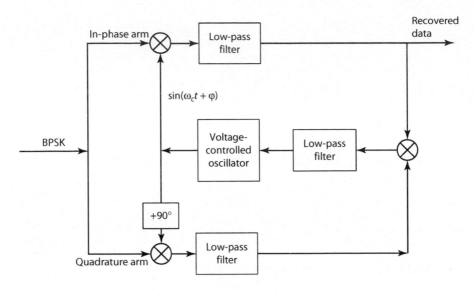

FIGURE 3.18   The Costas loop method of recovering the data in a BPSK system.

As can be seen from Figure 3.18, there are two arms to the Costas loop – in-phase and quadrature. To analyse the loop, we will take the BPSK to be

$$\text{BPSK} = m(t)\sin\omega_c t$$

where $m(t)$ is the data +1 or −1. We take the loop to be at the correct frequency but with a phase shift for the moment. After filtering, the in-phase arm gives

$$m(t)\sin\omega_c t \; . \; \sin\left(\omega_c t + \varphi\right)$$

$$= \frac{1}{2}m(t)\cos\varphi \tag{3.22}$$

which is the demodulated data if the loop is phase locked so that the cosine term is unity. The quadrature arm is used to correct for phase errors. Thus, after filtering,

$$m(t)\sin\omega_c t \; . \; \sin\left(\omega_c t + \varphi + \frac{\pi}{2}\right)$$

$$= \frac{1}{2}m(t)\cos\left(\varphi + \frac{\pi}{2}\right) \tag{3.23}$$

This is mixed with the demodulated signal (Equation 3.22) to give

$$\frac{1}{2}m(t)\cos\varphi\frac{1}{2}m(t)\cos\left(\varphi + \frac{\pi}{2}\right)$$

$$= -\frac{1}{8}m^2(t)\sin 2\varphi \tag{3.24}$$

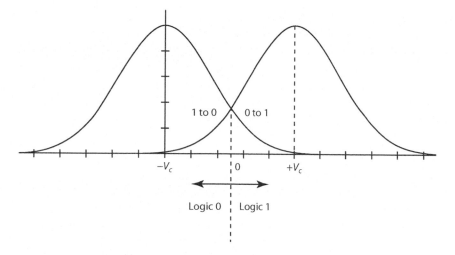

FIGURE 3.19   Probability density function for BPSK.

As the modulation is +1 or −1, the signal presented to the voltage-controlled oscillator (VCO) is proportional to sin 2φ. The loop operates to reduce this phase error to zero.

As regards the error performance of BPSK, we follow a similar analysis to that used for ASK and FSK. Thus, the pdf for BPSK is as shown in Figure 3.19. The threshold level is at 0 V and the difference between the threshold and the logic levels is V_c, the same as FSK. However, unlike FSK, there is only one decoding path and so there is only one lot of noise.

$$P_e = \frac{1}{2}\left(1 - \mathrm{erf}\left(\frac{V_c}{\sqrt{2\sigma^2}}\right)\right) \tag{3.25}$$

Comparing with the previous example, the carrier voltage can be reduced to 250 mV for the same amount of noise and the same probability of error. So, ASK required 500 mV, FSK required 354 mV and PSK needed 250 mV. Expressed in decibels (20 log as we are dealing with voltages), we gain 3 dB in using FSK (but with a high bandwidth) and 6 dB using PSK over ASK.

3.3.4   MATCHED FILTERING

When we considered baseband signalling in Section 3.2, we came across the need to filter signals to reduce the effects of noise. At that time, we considered raised-cosine filtering. There is an alternative that is the optimum digital filter – the matched filter. The matched filter is one that maximises the S/N regardless of the shape of the signal after filtering. It should be obvious that such a scheme is only applicable to digital transmission where we only need to decide if the signal is a one or a zero, present or not.

Consider a pulse, $p(t)$, with an FT of $P(\omega)$ filtered by a filter with a transfer function of $H(\omega)$. The output pulse, $p_o(t)$, will be given by the inverse FT:

$$p_o(t) = \frac{1}{2\pi} \int_{-\infty}^{\infty} P(\omega) H(\omega) e^{j\omega t} d\omega \tag{3.26}$$

The normalised signal power at a sampling time of t_o is

$$S = \left| \frac{1}{2\pi} \int_{-\infty}^{\infty} P(\omega) H(\omega) e^{j\omega t_o} d\omega \right|^2 \tag{3.27}$$

As the limits of integration in Equation 3.27 are $\pm\infty$, the noise must be double-sided. This is presented at the input to the filter and so the noise power at the output of the filter is

$$N = \frac{\eta}{2} \frac{1}{2\pi} \int_{-\infty}^{\infty} |H(\omega)|^2 d\omega \tag{3.28}$$

The S/N is simply the ratio of Equation 3.27 to Equation 3.28:

$$\frac{S}{N} = \frac{\left| \int_{-\infty}^{\infty} P(\omega) H(\omega) e^{j\omega t_o} d\omega \right|^2}{\frac{\eta}{2} \int_{-\infty}^{\infty} |H(\omega)|^2 d\omega} \tag{3.29}$$

What we now need to find is the maximum S/N. To do this, we need to use the Schwartz inequality:

$$\int_{-\infty}^{+\infty} f_1^*(x) f_1(x) dx \int_{-\infty}^{+\infty} f_2^*(x) f_2(x) dx \geq \left| \int_{-\infty}^{+\infty} f_1^*(x) f_2(x) dx \right|^2 \tag{3.30}$$

The equality only holds if

$$f_1 = K f_2^*(x) \tag{3.31}$$

Equating $f_1(x)$ to $P(\omega)$ and $f_2(x)$ to $H(\omega)$ exp $j\omega t_o$, after some re-arranging Equation 3.30 becomes

$$\int_{-\infty}^{+\infty} |P(\omega)|^2 d\omega \geq \frac{\left| \int_{-\infty}^{+\infty} P(\omega) H(\omega) e^{j\omega t_o} d\omega \right|^2}{\int_{-\infty}^{+\infty} |H(\omega)|^2 d\omega} \tag{3.32}$$

This result can be substituted into Equation 3.29 to give

$$\int_{-\infty}^{+\infty} |P(\omega)|^2 d\omega \frac{2}{\eta} \geq \frac{S}{N} \tag{3.33}$$

The integral term is the energy in the signal, E, and the maximum S/N occurs with the equality satisfied, and so

$$\frac{S}{N} = \frac{2E}{\eta} \tag{3.34}$$

This equation shows that the signal power is E and so the signal amplitude is $\sqrt{E}$. It also shows that the S/N does not depend on the shape of the received signal but only on the energy in the pulse. The filter transfer function can be found as follows. From Equation 3.31:

$$P(f) = KH(f)e^{j\omega t_o}$$

Therefore,

$$H(f) = \frac{1}{K}P(f)e^{-j\omega t_o} \tag{3.35}$$

where K is a constant of proportionality, which we will set to one. The impulse response of the filter is therefore

$$h(t) = \int_{-\infty}^{+\infty} P(\omega)e^{-j\omega t_o}e^{j\omega t}d\omega$$

and so

$$h(t) = p(t_o - t) \tag{3.36}$$

This shows that a matched filter has an impulse response that is the time reversal of the received pulse. For a rectangular pulse this equates to an integrate and dump filter as shown in Figure 3.20. The signal and noise are integrated by the resistor-capacitor (RC) time constant and the output is sampled at the end of the bit time, T. After sampling, the charge on the capacitor must be removed, or dumped, ready for the next pulse to be integrated. There are two problems with this circuit: the RC time constant gives an exponential rise which is not an integral function; and it is impera-tive that the dump function completely removes the charge otherwise there will be intersymbol interference. It is possible to use an integrator to recover the baseband if speed is not an issue. Indeed, it is possible to use a software solution based on digital signal processing (DSP).

The pulse at the output of the integrator is found by substituting Equation 3.35 into Equation 3.26 to give

$$p_o(t_o) = \frac{1}{2\pi}\int_{-\infty}^{\infty} P(\omega)P(\omega)d\omega = E \tag{3.37}$$

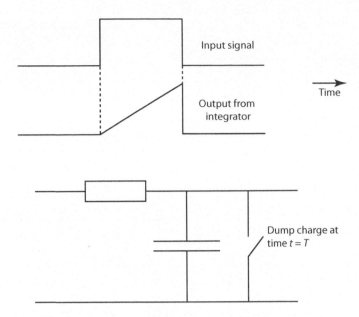

FIGURE 3.20   An integrate and dump matched filter: (a) ideal response and (b) an approximate solution.

The noise at the output of the integrator is found by substituting Equation 3.37 into Equation 3.34 to give

$$\frac{S}{N} = \frac{E^2}{N} = \frac{2E}{\eta}$$

And so,

$$N = E\frac{\eta}{2} \tag{3.38}$$

3.3.5   ORTHOGONAL FREQUENCY-DIVISION MULTIPLEXING

Our work on ASK, FSK and PSK has shown us that the S/N can be quite low and still we get good error performance. We will see in Section 4.1 that the S/N for analogue signals can be quite high (10^6) and so there is a lot to be gained if we use digital techniques for broadcasting – more sensitive receivers implies a lower error rate or greater transmission distance.

Consider a single carrier transmitting at a frequency of 100 MHz. Let this carrier be amplitude modulated by a 10 Mbit/s data stream to a depth of 100% so that the carrier is fully extinguished for a logic zero. From Section 2.7, we know that the spectrum will have nulls at harmonics of 10 MHz either side of the carrier (90 and 110 MHz, 80 and 120 MHz, etc.). As also noted in Section 2.7, we do not need an infinite bandwidth in the transmitter or receiver because it is the high frequencies in the spectrum that give fast rise and fall times and these can be sacrificed to minimise the noise. In general, we can use 0.7 times the bit rate so the bandwidth would be 2×7 MHz centred on 100 MHz. (The choice of 0.7 times the bit rate is a compromise between maintaining pulse shape and minimising noise. In effect, it is an approximation to the full raised-cosine filter.)

The advantage of modulating a single frequency is that it is very simple to implement and demodulate. The big disadvantage is that it is very susceptible to multipath interference. Consider the 10 Mbit/s data just described; the bit time is $1/10 \mu s = 0.1$ μs. Take two transmission paths and say that the time delay between them is longer than the bit time. This situation is shown in Figure 3.21a. The delay will cause corruption of bit b_1 by the delayed bit b_2 resulting in detection errors. As well as the time delay between bits, there will also be a phase shift in the carrier and this could cause further difficulties. If the time delay is less than the bit time, we have the situation shown in Figure 3.21b. In this case, there is a guard interval in which interference occurs, then there is a time in which the pulse is clear of interference. It is during this time that detection takes place. In digital audio broadcasting (DAB), the guard is ¼ the data time, whereas in DBT-Terrestrial (DVB-T) it varies from ¼ to 1/32. Taking our bit rate of 10 Mbit/s, the bit time is 100 ns and so the guard interval is 25 ns. This corresponds to a path length of 7.5 km. If the bit rate is 1/100 of the 10 Mbit/s (done by using parallel transmission, which is considered next), the path difference becomes 750 km. As the signal strength decreases with distance from the transmitter, there is unlikely to be any catastrophic interference. Using this system, it is possible to have transmitters using the same frequency in close proximity and so the power in data symbols will add and the signal strength will increase. There is a problem with multipath distortion in that it destroys orthogonality, a very important signal property.

If we have a directional antenna at both ends of a link, we might be able to eliminate multipath interference. However, the advantages of digital transmission are a better S/N performance and better quality and so it would be useful in mobile links where signal strengths are variable and multipath interference is present. This is where the guard interval is useful. This guard interval is not kept empty; instead, a

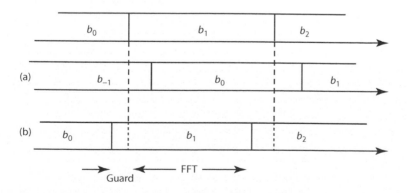

FIGURE 3.21 Multipath distortion on a single frequency carrier.

portion of the end of the symbol is copied into the guard to maintain orthogonality as we will see shortly.

Before we begin our studies of digital transmission systems, let us return to the Shannon–Hartley theorem (Section 1.6), which is reproduced here as

$$C = B \log_2 S / N \qquad (3.39)$$

where:

C is the channel capacity in bit per second
B is the bandwidth of the channel
S/N is the signal to noise ratio

This equation shows that if the bandwidth of the channel is increased, the capacity will also increase. This should be self-evident – a telephone line will not carry 1 Gbit/s data but a high bandwidth optical link will. Also, if the S/N is high (either a large signal or small noise), the potential capacity will be higher. If $\log_2$ S/N is 1, the capacity is proportional to the bandwidth and so an increased bandwidth would give an increased capacity. However, this does not take account of the error rate. A $\log_2$ S/N of 1 would result in a large number of errors, which could be reduced by introducing error correction. If we take a 4 kHz bandwidth (that of a telephone line) and an S/N of 25 dB (316 as a ratio), we get a capacity of 33 kbit/s. If the bandwidth of the channel is increased, as in asynchronous digital subscriber links (ADSL), the capacity will also increase.

The problem of multipath distortion is one reason why single frequency transmission is not very popular for high data rates. It is a question of the pulse time being small relative to the time delay. A solution is to spread the data transmission over a large number of carriers. So, the 10 Mbit/s data stream can be spread over 100 different carriers each operating with 10/100 Mbit/s or 100 kbit/s. With this data rate, the pulse time is 100 times greater at 10 µs and so the amount of delay that can be accommodated is 100 times greater. Figure 3.22 shows the demultiplexing of a data stream into four lines. As can be seen, the demultiplexing into four data lines has reduced the bit rate by a factor of 4.

These four data lines can easily be used to modulate four separate carriers such as shown in Figure 3.23. There is an immediate problem; the tails and precursors on the carriers are interfering with each other. The solution is to space the carriers out even further. This has the effect of increasing the overall bandwidth to such an extent that it may be greater than operating with a single data stream. Fortunately, it is possible to space the carriers so that there is no interference between the channels. Such a scheme is known as orthogonal frequency-division multiplexing (OFDM). Orthogonal simply means that there is no interference between carriers. (Two vectors are orthogonal if they do not affect each other. An example is a vector acting along the x axis that does not affect another vector acting along the y axis. They are orthogonal to each other.)

Orthogonality can be produced by using harmonics of a low-frequency fundamental that has a period equal to the bit time (Figure 3.24). So, the 10 Mbit/s data

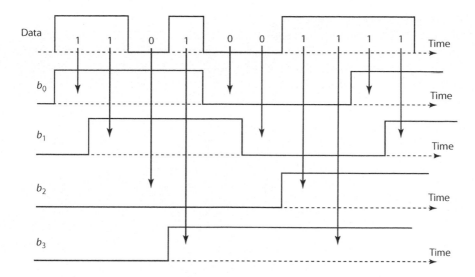

FIGURE 3.22 Demultiplexing of data into four channels.

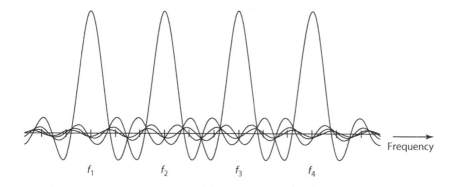

FIGURE 3.23 Frequency-division multiplexing of four data streams onto four carriers.

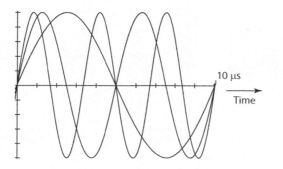

FIGURE 3.24 Orthogonal harmonic sine waves over the bit time.

demultiplexed down to 100 kbit/s gives a fundamental of 100 kHz and harmonics at 200 kHz, 300 kHz, etc. This is still some way from the 100 MHz signals in use; however, as we have seen with analogue transmitters, frequency shifting can get the multiplex to the final frequency. The spectrum of OFDM showing seven modulated carriers is shown in Figure 3.25.

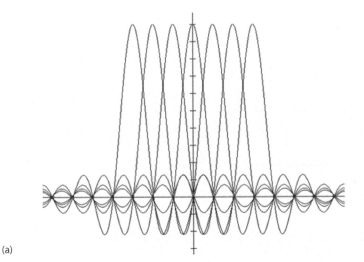

(a)

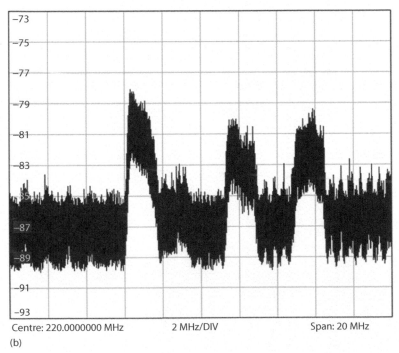

Centre: 220.0000000 MHz   2 MHz/DIV   Span: 20 MHz
(b)

FIGURE 3.25   (a) Spectrum of orthogonal carriers and (b) measured spectrum of a digital audio broadcast.

Modulation of the carriers is done in the usual fashion by mixing a carrier with the data. Demodulation consists of mixing signals down to the baseband and integrating over the time period of the data (10 μs). Consider the fundamental signal being mixed with a local oscillator, v_{LO}, to the baseband.

$$v_{OFDM}(t) = m_1(t)\cos\omega t + m_2(t)\cos 2\omega t + m_3(t)\cos 3\omega t + \text{etc.}$$

$$v_{LO}(t) = \cos\omega t$$

Therefore,

$$v_{OFDM}(t)v_{LO}(t) = m_1(t)\cos^2\omega t + m_2(t)\cos\omega t\cos 2\omega t + m_3(t)\cos\omega t\cos 3\omega t + \text{etc.}$$

This gives terms in

Baseband	½ $m_1(t)$	
Fundamental	½ $m_2(t)$	
2nd harmonic	½ $m_1(t)$	½ $m_3(t)$
3rd harmonic	½ $m_2(t)$	
etc.		

We require the baseband to recover $m_1(t)$. The baseband appears as zero frequency (dc) and, as the frequency terms are all completely periodic over the bit time (10 μs), their integrals over the bit time are zero. So, by simply integrating over the bit time (an integrate and dump), we will only get the channel we require. The same argument applies to the other carriers.

Figure 3.26 shows a functional block diagram of an OFDM modulator and demodulator. A major disadvantage of OFDM is immediately apparent: there are N mixers and N orthogonal oscillators in the transmitter and the same in the receiver, in addition to N integrators. This is a lot of circuitry. Fortunately, the operations we are performing in the transmitter are that of an inverse fast Fourier transform (IFFT) and in the receiver the fast Fourier transform (FFT). For the transmitter, the IFFT will furnish us with the orthogonal carriers (Appendix VIII). This signal processing takes place at a very low frequency and is well within the range of commercial DSP chips.

3.3.6 QUADRATURE AMPLITUDE MODULATION

It is possible to combine ASK with PSK to give quadrature amplitude modulation (QAM). This is used in Europe with OFDM to carry digital TV and radio in both Europe and the United States. Let us examine 4-PSK, also known as quadrature PSK (QPSK), shown in Figure 3.27. As can be seen, there are four symbols occupying four equally spaced phase shifts. Four symbols means that two bits of data can be encoded – 00, 01, 10, 11 – and so the data rate can be increased by a factor of 2 for the same bandwidth. Errors generally occur when noise causes a symbol to rotate

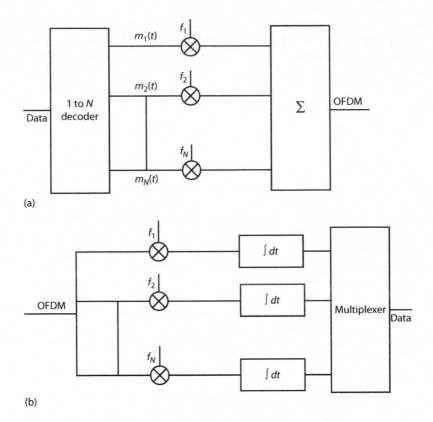

(a)

(b)

FIGURE 3.26 (a) An OFDM modulator and (b) an OFDM demodulator.

into an adjacent symbol space. This is because the distance between opposite symbols, 01 and 10 for instance, is greater than the distance between adjacent symbols.
Thus, noise will affect adjacent symbols first. The symbol phases are allocated using
the Gray code to minimise the error rate. (01 could go to 00 giving 1 error or go to
11 giving 1 error. It is unlikely to go to 10 for the reasons just given.) Also shown in

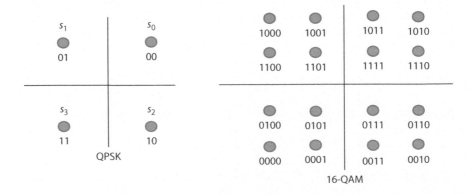

FIGURE 3.27 Constellation diagrams for QPSK and 16-QAM.

Figure 3.27 is the constellation for 16-QAM. As can be seen, changes in amplitude have been added so that four bits of data can be transmitted. Also note that the codes have been distributed so that an error to an adjacent symbol generates one bit in error. There are higher levels in use such as 64-QAM and even 256-QAM. Such high constellations are more susceptible to noise and error, but error correction can be used to maximise performance. One thing worth noting is that the peak-to-average power ratio (PAPR) can be very large for large constellations and this causes difficulties when designing power amplifiers.

We will now determine the error rate of QPSK. As just noted, errors are likely to occur between adjacent symbols not diametrically opposite ones. Thus, the bit error rate is the same as the symbol error rate. We will assume additive white Gaussian noise and a decoder as shown in Figure 3.28.

As shown in Figure 3.27, we have four signals: $\pm A \cos(\omega t + \pi/4)$ and $\pm A \cos(\omega t - \pi/4)$. Consider the symbol pair $\pm A \cos(\omega t + \pi/4)$. This is mixed with $\cos \omega t$ to give

$$\pm A\cos\left(\omega t + \frac{\pi}{4}\right)\cos \omega t = \pm \frac{A}{2}\left(\cos \frac{\pi}{4} + \cos\left(2\omega t + \frac{\pi}{4}\right)\right) \qquad (3.40)$$

The second harmonic is filtered out to leave

$$\pm \frac{A}{2\sqrt{2}} \qquad (3.41)$$

As we have already seen, the factor of ½ comes about from the mixing of the signal with the carrier and the same factor of ½ applies to the noise as well. So, we will ignore this factor, giving

$$\pm \frac{A}{\sqrt{2}} \qquad (3.42)$$

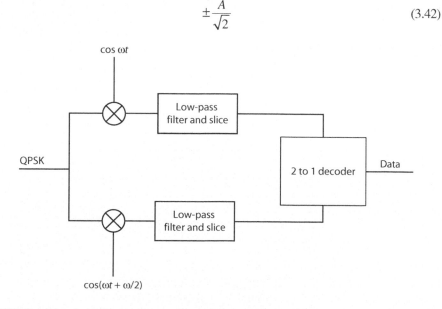

FIGURE 3.28 A QPSK decoder.

This is the amplitude of all four symbols: s_0, s_1, s_2 and s_3. A decision is then made in the usual fashion. The noise is Gaussian in form and so the treatment is similar to that in Section 3.3.3 with the pdfs centred on $\pm\dfrac{A}{\sqrt{2}}$. So,

$$P_{es} = \int_{A/\sqrt{2}}^{+\infty} \frac{1}{\sqrt{2\pi}\sigma} \exp\!\left(-v^2\!\Big/2\sigma^2\right) dv \tag{3.43}$$

Substituting

$$u^2 = v^2\!\Big/2\sigma^2 \quad u = v\!\Big/\sqrt{2}\sigma \quad du = dv\!\Big/\sqrt{2}\sigma$$

gives

$$P_{es} = \int_{A/2\sigma}^{+\infty} \frac{1}{\sqrt{\pi}} \exp\!\left(-u^2\right) du$$

$$= \frac{1}{2}\left(1 - \operatorname{erf}\left(A\!\Big/2\sigma\right)\right) \tag{3.44}$$

This is the error probability for one arm of the decoder in Figure 3.28. An identical result applies to the other decoding arm. The probability of correctly decoding the data is the product of the two arms correctly decoding the data. As the symbol error is the same for both arms:

$$P_{\text{correct}} = \left(1 - P_{es}\right)\left(1 - P_{es}\right)$$

$$= 1 - 2P_{es} + P_{es}^2$$

The error probability is usually small and so P_{es}^2 is even smaller and can be neglected. The probability of an error, P_e, is given by

$$P_e = 1 - P_{\text{correct}}$$

$$= 2P_{es} \tag{3.45}$$

$$= \operatorname{erfc}\left(\frac{A}{2\sigma}\right)$$

If a matched filter is used, A is replaced by $\sqrt{E}$ and σ^2 by $\eta/2$ to give

$$P_e = \left(1 - \operatorname{erf}\left(\frac{1}{\sqrt{2}}\sqrt{\frac{E}{\eta}}\right)\right) \tag{3.46}$$

The error rate of QAM is higher than BPSK. However, the data rate is doubled for the same signal bandwidth. This is of importance when considering bandwidth-limited channels – near enough all channels these days. What we have done in effect is increased the capacity of the channel by using QAM but at the expense of a higher

S/N. This is shown in Shannon's law (Equation 3.39). This is important if we want to send data over a limited bandwidth – we can use multilevel signalling which requires a larger S/N. The effect of an increased error rate can be reduced by using various coding techniques examined next.

3.4 CODING

Coding is used to increase the reliability of transmission by adding a certain number of redundant bits. It can be very simple as in parity checking, or more complex as in Reed–Solomon (RS) coding. We will examine several different coding schemes that go by the collective name of forward error correction (FEC) where the forward refers to the forward data path from transmitter (source) to the receiver.

3.4.1 Parity Check

Parity is the simplest form of error checking. It does not give us the power to perform error correction though. To explain parity, it is best to look at some examples (Table 3.1). Table 3.1 lists some sample 7-bit words much the same as would be generated by the ADC of Section 3.1.

If an error occurs – a logic [0] to a [1] or a logic [1] to a [0] – the parity check will be wrong and an error can be detected. The receiver can then request that the transmitter re-sends the faulty word. However, if two errors occur, the parity check will be satisfied and no error will be detected. Also, there is no way of correcting the error.

Mathematically, even parity can be expressed as

$$b_1 \oplus b_2 \oplus b_3 \oplus b_4 \oplus b_5 \oplus b_6 \oplus b_7 \oplus p_1 = 0 \qquad (3.47)$$

where $\oplus$ is modulo 2 addition – an EXOR function. Odd parity is the same expression except that the resultant is equal to one.

3.4.2 Hamming Code

One way to correct errors is to use the Hamming code. Before we go on to discuss this technique in detail, we need to define a number of terms. A block code (n, k) has 2^k data words and $n-k$ parity bits. Thus, a 7, 4 code has 2^7 7-bit codewords, 2^4 4-bit

TABLE 3.1

Even Parity on 7-Bit Words

b_1	b_2	b_3	b_4	b_5	b_6	b_7	P_1 (even)
0	0	0	1	1	1	0	1
0	0	0	0	0	0	0	0
1	1	1	1	0	1	1	0
1	1	1	0	0	0	0	1

data words and 3 bits of parity. A code is said to be linear if the EXOR of two code-words is another valid codeword and the EXOR of a codeword with itself is zero (this should be self-evident). The Hamming weight of a codeword is the number of [1]s in the codeword. So, the word 100111 has a Hamming weight of 4. The Hamming distance is the number of places two codewords differ. Thus, the Hamming distance for 110001 and 100000 is found by taking the EXOR to give 010001 or $d=2$. If s errors can be detected and t errors corrected ($s \geq t$), the following holds true:

$$d \geq 2t + 1 \tag{3.48}$$

Note that there has to be some redundancy in the data words if the minimum Hamming distance is to be 3. (Use of Equation 3.48 shows that one error can be corrected.)

Let us examine two codes, one of which is linear (Table 3.2). The even parity code is linear because a linear code must contain the all-zero condition. Note also that the minimum Hamming distance is 2, so no errors can be corrected.

To return to the Hamming code, the (7, 4) code is widely used and to analyse it we must follow some basic rules. Figure 3.29 shows the basic Hamming word. The P represents parity bits and the D the data bits. As can be seen, it is a 7-bit word with 4 bits of data. The position of the parity bits follows the powers of 2; thus $2^0 = 1$, $2^1 = 2$, $2^2 = 4$, etc. In order to generate the parity bits, the following applies: P_1 takes positions 1, 3, 5, 7; P_2 takes 2, 3, 6, 7; P_4 takes 4, 5, 6, 7. The bit is either set or reset according to even parity. Let us consider an example data word of 1001. Figure 3.29 becomes Figure 3.30.

It is possible to detect and correct a bit error. Consider bit 6 to change from 0 to 1 due to noise. To detect the error, the parity bits for the received word are calculated (Figure 3.31). The location of the error is $2+4=D_6$ and so the bit needs to be corrected.

TABLE 3.2
A Linear and Non-Linear Code with Parity Bits (Least Significant Bit)

Even Parity		Odd Parity	
0000	00000	0000	00001
0001	00011	0001	00010
0010	00101	0010	00100
0011	00110	0011	00111
0100	01001	0100	01000
0101	01010	0101	01011
0110	01100	0110	01101
0111	01111	0111	01110
1000	10001	1000	10000
etc.	etc.	etc.	etc.

$$P_1\ P_2\ D_3\ P_4\ D_5\ D_6\ D_7$$

FIGURE 3.29 A 7-bit Hamming word.

0 0 1 1 0 0 1

$P_1 = ?\,1\,0\,1 = 0$ even parity

$P_2 = ?\,1\,0\,1 = 0$ even parity

$P_4 = ?\,0\,0\,1 = 1$ even parity

FIGURE 3.30 An even parity (7, 4) Hamming word.

0 0 1 1 0 1 1

$P_1 = ?\,1\,0\,1 = 0$ even parity

$P_2 = ?\,1\,1\,1 = 1$ sent as 0

$P_4 = ?\,0\,1\,1 = 0$ sent as 1

FIGURE 3.31 An error in even parity (7, 4) Hamming word.

3.4.3 CYCLIC REDUNDANCY CODE

The next code we will examine is the cyclic redundancy code (CRC). This code gives an indication of an error but will not correct it. It works by "dividing" a data sequence by a divisor and computing the remainder – the CRC. Although the action appears to be division, it isn't as the EXOR function is used. An example will help.

Consider the sequence 10110111 ($x^7 + x^5 + x^4 + x^2 + x^1 + x^0$) and the divisor 101 ($x^2 + x^0$). The divisor is 3 bits long and so two zeroes must be appended to the sequence. The resultant is then divided by the divisor. To form the codeword, the CRC bits (2 off) replace the two zeroes in the data word. Thus, 1011011100 becomes 1011011111. Decoding follows a similar process in that the received word is divided by the same divisor. If the remainder is zero, then there are no errors in the received word (Figure 3.32). If errors are detected, a request to re-send the data can be sent to the transmitter and the word can be sent again.

3.4.4 CONVOLUTION CODING, MAXIMUM LIKELIHOOD AND VITERBI DECODING

In convolutional coding, the input data is gated with itself to give an additional code containing extra bits. Convolutional codes are referred to as (n, k, m), where n is the number of output bits, k is the number of input bits and m is the number of registers used. As an example, let us consider the code (2, 1, 4), shown in schematic form in Figure 3.33. This has two output bits, one input bit and a 4-bit shift register.

The impulse response of the coder for a logic one is obtained by passing a one through the coder, and the impulse response for a logic zero comes about from passing a logic zero through the coder. As a logic one propagates through the coder, b_0 will always be 1 because d_0 to d_3 will hold a logic one. Similarly, when the 1 is

Encoder   Decoder

```
        Encoder                        Decoder
        1000100011                     10010011
101)1011011100                101)1011011111
     101                            101
     0001                           0001
       000                            000
       0010                           0010
         000                            000
         0101                           0101
           101                            101
           0001                           0001
             000                            000
             0011                           0011
               000                            000
               0110                           0111
                 101                            101
                 0110                           0101
                   101                            101
                   011                            000
```

Here the remainder is 11   Here the remainder is zero

FIGURE 3.32   Cyclic redundancy check of an 8-bit word.

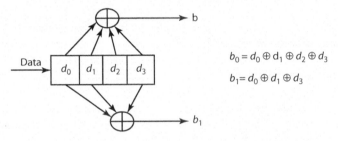

$$b_0 = d_0 \oplus d_1 \oplus d_2 \oplus d_3$$

$$b_1 = d_0 \oplus d_1 \oplus d_3$$

FIGURE 3.33   A convolutional coder.

clocked through to d_2, b_1 will be zero, but 1 otherwise (Table 3.3). Note that the single logic one has generated eight bits of data as it is clocked through – 11 11 10 11.

To find the response to the input data, we need to convolve the data with the respective impulse response. Let us try 1010 as an input. Table 3.4 shows how the output is obtained. The result is the same as that obtained by passing 1010 through the coder in Figure 3.33. The various outputs for all 16 inputs can be found in a similar fashion.

TABLE 3.3

Impulse Response of Convolutional Coder Shown in Figure 3.33

Clock	d_0	d_1	d_2	d_3	b_0	b_1
1	1	0	0	0	1	1
2	0	1	0	0	1	1
3	0	0	1	0	1	0
4	0	0	0	1	1	1

TABLE 3.4
Output of Convolutional Coder of Figure 3.33 for 1010 as the Input

Data				b_0b_1	b_0b_1	b_0b_1	b_0b_1	b_0b_1	b_0b_1	b_0b_1
1				11	11	10	11			
	0				00	00	00	00		
		1				11	11	10	11	
			0				00	00	00	00
Output				11	11	01	00	10	11	00

Decoding of the data can be achieved by a variety of means. The first we consider uses a maximum likelihood scheme in which the corrupted data is compared to all possible codewords. Table 3.5 details the coded data for all 16 possible inputs for the convolutional coder of Figure 3.33. The corrupted, received codeword is compared to all possible codewords and the number of bits not in agreement is computed. This is known as the Hamming distance. The codeword with the lowest Hamming distance (the most bits in agreement) is the maximum likelihood codeword. It is clear from Table 3.5 which code was transmitted. So, the use of convolutional codes with a maximum likelihood detector allows us to correct some detection errors. It should be remembered that this is at the expense of an increased data rate (14 bits transmitted for 4 bits of data).

TABLE 3.5
Maximum Likelihood Decoding Table for the Convolution Coder of Figure 3.33

Data	Coded	Received Codeword	Hamming Distance
0000	00 00 00 00 00 00 00	00 11 00 00 10 01 11	6
0001	00 00 00 11 11 10 11	00 11 00 00 10 01 11	7
0010	00 00 11 11 10 11 00	00 11 00 00 10 01 11	9
0011	00 00 11 00 01 01 11	00 11 00 00 10 01 11	6
0100	00 11 11 10 11 00 00	00 11 00 00 10 01 11	7
0101	00 11 11 01 00 10 11	00 11 00 00 10 01 11	6
0110	00 11 00 01 01 11 00	00 11 00 00 10 01 11	6
0111	**00 11 00 10 10 01 11**	**00 11 00 00 10 01 11**	**1**
1000	11 11 10 11 00 00 00	00 11 00 00 10 01 11	9
1001	11 11 10 00 11 10 11	00 11 00 00 10 01 11	6
1010	11 11 01 00 10 11 00	00 11 00 00 10 01 11	6
1011	11 11 01 11 01 01 11	00 11 00 00 10 01 11	7
1100	11 00 01 01 11 00 00	00 11 00 00 10 01 11	10
1101	11 00 01 10 00 10 11	00 11 00 00 10 01 11	9
1110	11 00 10 10 01 11 00	00 11 00 00 10 01 11	11
1111	11 00 10 01 10 01 11	00 11 00 00 10 01 11	6

The maximum likelihood decoder presented in Table 3.5 is quite resource intensive in that the received codeword is compared to every possible codeword generated by the convolutional coder. An alternative technique is the Viterbi decoder which acts on data as it arrives.

Consider the received sequence 00 11 00 01 01 11 00 taken from Table 3.5. Reception of the first 00 puts the decoder in the first half of Table 3.5, thus limiting the choice to eight possible codewords. Reception of the next two bits (11) limits the choice to four codewords. The following 00 bits gives a choice of two codewords, and finally 01 gives 0110 as the decoded word. Now consider what happens when there is an error so that 00 11 01 01 01 11 00 is received. As before, the initial 00 places the decoder in the first half of Table 3.5. The next two bits (11) give four codewords as before. Now comes the error bit 01. This is an invalid set of bits – only 11 or 00 are allowed and, assuming only one bit is in error, 01 could have been 11 or 00. Thus, the choice has not been narrowed down. The next pulses are 01 resulting in two possible codewords 0101 and 0110. The next 01 gives the codeword 0110 which is correct. So an error in the data stream has been corrected.

An alternative representation is the lattice diagram. Rather than produce a lattice for the 4-bit code we have just examined, let us work with the code generated by the coder of Figure 3.34. (It is left to the reader to complete the lattice diagram for the 4-bit code we have been considering.) Table 3.6 shows the output of the convolutional coder as a zero (most significant bit [MSB]) and then a one (MSB) is clocked through. This can be represented on a lattice diagram as shown in Figure 3.35. Here, the first register, d_0, is not used as we can take it to be the input data. It should be noted that this lattice diagram shows the contents of the shift registers and not the output of the coder. Also, it is worth noting that if a zero is clocked through from the first register, the state change arrow is pointing up. The arrows point down for a logic one being clocked through. As an example, consider the register state to be 10.

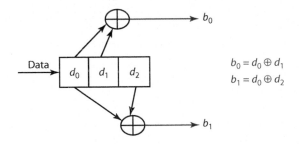

$$b_0 = d_0 \oplus d_1$$
$$b_1 = d_0 \oplus d_2$$

FIGURE 3.34 A 1/2-bit convolutional coder.

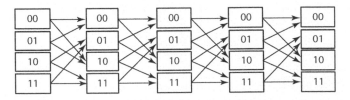

FIGURE 3.35 Lattice diagram of the registers in a convolutional coder.

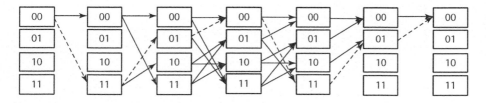

FIGURE 3.36 Viterbi lattice decoder.

A logic zero clocked in will give 01 and so there is an upward arrow. A logic one being clocked in will give 11 and a downward pointing arrow.

A trellis is also used in the demodulation of a convolutional code. This is called *Viterbi decoding* and it uses a lattice as shown in Figure 3.36. Figure 3.36 is a lattice diagram for the sequences generated by the coder of Figure 3.34 and the outputs of Table 3.6. It shows all possible valid combinations of the 2-bit words that make up the output. As an example, the dotted line shows a path through the lattice beginning at 00, then progressing to 11, 01, 00, 11, 01, 00 which corresponds to 111 from Table 3.6. The power of this decoding scheme lies in what happens when there is an error. Let us take the dotted path as before, but have a sequence 00 11 01 01 (error). We know that there is an error because a sequence with two bits in error is not allowed. So, we go back to the previous 2-bit word 01. There are two options – 00 or 11. At this point, both are equally valid so let us take 11. The next received sequence is 11, which is not allowed if the previous bits were 11. So, this is a false trail and we should have used 00. Thus, we have corrected a received word with an error in it.

3.4.5 REED–SOLOMON CODING

A further error correction scheme is used in digital broadcasting – Reed–Solomon coding. This relies on taking blocks of data and adding a number of parity blocks at the end of each set of data symbols, as shown in Figure 3.37. RS codes are referred to as RS(n, k) with s bit symbols. An example is RS(255, 247) $s = 8$. So, the symbols

TABLE 3.6
Output of the Convolutional Coder of Figure 3.34 as a Zero (MSB) and Then a One (MSB) Is Clocked In

Data				$b_0 b_1$			
0 00	00	00	00	00	00	00	00
0 01	00	11	10	01	00	00	00
0 10	00	00	11	10	01	00	00
0 11	00	11	01	11	01	00	00
1 00	00	00	00	11	10	01	00
1 01	00	11	10	10	10	01	00
1 10	00	00	11	01	11	01	00
1 11	00	11	01	00	11	01	00

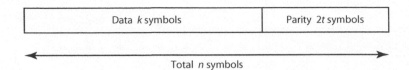

FIGURE 3.37   A Reed–Solomon data word.

are 8 bits long, 255 symbols are used with $2t=8$ parity symbols. With $2t=8$, $t=4$ and that is the number of symbols (8-bit words) that the RS code can correct. Use of RS and other error correction techniques can reduce the error rate considerably. For instance, a link error rate of 1 in 10^3 can become 1 in 10^9 with error correction. This means that lower received power is required to get a good error rate.

Our number system is an example of an infinite field – it has no end as there is always a number greater than infinity. We can convert into a finite field by populating the field with the units from our infinite field. Thus, we can have 0, 1, 2, 3, 4, 5, 6, 7, 8, 9, 0, 1, 2, etc., within the field. Note that the powers of 10 are ignored; we only populate the field with remainders. RS codes use the finite fields known as *Galois fields* (GF), which have some rather interesting arithmetic. As an example, we will consider the GF(2^3) or GF(8) field. This field will have eight elements to it and it can be populated by the field generator polynomial $p(\alpha)$ given by

$$\alpha^3 + \alpha + 1 \tag{3.49}$$

The value of α can be set to any prime number but we will use 2 because we are using binary digits. In our GF, addition is done using an EXOR function, so $1+1=0$ and $1-1=0$. We can say $1+1=1-1=0$, and that addition is the same as subtraction. This is a very important relationship that we will use extensively very soon. We can set α as a root of Equation 3.48 to give

$$\alpha^3 + \alpha + 1 = 0$$

which yields

$$\alpha^3 = \alpha + 1 \tag{3.50}$$

where we have used the fact that subtraction is the same as addition in our GF(8) field (Table 3.7). We can populate the field as follows, but remember that α is 2. The field is populated by successive multiplications by α. Thus, the entry for $\alpha + 1$ becomes $\alpha^2 + \alpha$, and we also use the result of Equation 3.49. One thing worth noting is that $\alpha + \alpha$ equals 0. This is a direct result of the EXOR addition we are applying and so the addition of identical field elements results in zero.

To produce the parity bits for the RS code, we need to be able to add and multiply elements. Appendix I shows the addition of elements from 0 to 7. Recall that addition is done in binary mod 2 (EXOR). As an example, $2+6=010+110=100$ or

TABLE 3.7
How GF(8) Is Populated

Power	Galois Field Polynomial	Decimal	Comment	Binary Code
0	0	0		000
α^0	α^0	1		001
α^1	α^1	2		010
α^2	α^2	4		100
α^3	$\alpha+1$	3	Definition of α^3 used	011
α^4	$\alpha^2+\alpha$	6		110
α^5	$\alpha^2+\alpha+1$	7	Definition of α^3 used	111
α^6	α^2+1	5	Definition of α^3 used	101
α^7	α^0	1	Repeating sequence	001

4 decimal. Another example is $3+5=0011+0101=0110=6$ decimal. For the multiplication table, a polynomial representation is used. Consider 3×6. In decimal the answer is 18, which is clearly beyond our base 8. However, 3 is $\alpha+1$ and 6 is $\alpha^2+\alpha$ and so 3×6 is 1 where use has been made of Equation 3.50.

We are now in a position to design an RS code. The one used for digital TV is (255, 239) and is too complicated to be analysed here. Instead, we will design a (7, 3) code that will correct a maximum of two symbols each with three bits. The starting point is a generator polynomial $g(x)$ such as

$$0g(x)=\left(x+\alpha^0\right)\left(x+\alpha^1\right)...\left(x+\alpha^{2t-1}\right)$$

$$g(x)=\left(x+\alpha^0\right)\left(x+\alpha^1\right)\left(x+\alpha^2\right)\left(x+\alpha^3\right)$$

$$=(x+1)(x+2)(x+4)(x+3)$$

$$=\left(x^2+3x+2\right)(x+4)(x+3) \tag{3.51}$$

$$=\left(x^3+7x^2+5x+3\right)(x+3)$$

$$=x^4+4x^3+7x^2+7x+5$$

In the derivation of this, we have used Equation 3.50 and the summation and multiplication tables of Appendix I. The next step is to divide a message polynomial, $m(x)$, by the generator polynomial Equation 3.51. Let the message be 110010001 and consider symbols that are 3 bits long. So, $m(x)$ is

$$m(x)=110\ 010\ 001$$

$$=6x^2+2x+1 \tag{3.52}$$

The parity bits must be added to this data word to make up the symbol. In our case, we have four symbols of parity and so the message word has to be shifted by four places to make room, i.e. $m(x)$ must be multiplied by x^4. Thus, Equation 3.52 becomes

$$m(x) = 6x^6 + 2x^5 + x^4 + 0x^3 + 0x^2 + 0x + 0 \qquad (3.53)$$

We now divide the message (Equation 3.53) by the generator (Equation 3.51) to produce the parity bits. The rules of multiplication and addition/subtraction are shown in Appendix I.

$$
\begin{array}{r}
6x^2 + 7x + 4 \\
x^4 + 4x^3 + 7x^2 + 7x + 5 \overline{)\, 6x^6 + 2x^5 + 1x^4 + 0x^3 + 0x^2 + 0x + 0} \\
\underline{6x^6 + 5x^5 + 4x^4 + 4x^3 + 3x^2} \\
7x^5 + 5x^4 + 4x^3 + 3x^2 + 0x \\
\underline{7x^5 + 1x^4 + 3x^3 + 3x^2 + 6x} \\
4x^4 + 7x^3 + 0x^2 + 6x + 0 \\
\underline{4x^4 + 6x^3 + 1x^2 + 1x + 2} \\
1x^3 + 1x^2 + 7x + 2
\end{array}
$$

The remainder is $1x^3 + 1x^2 + 7x + 2$, which translates to 001 001 111 010. These bits are appended to the message to give

$$m(x) = 6x^6 + 2x^5 + 1x^4 + 1x^3 + 1x^2 + 7x + 2 \qquad (3.54)$$

As a check, this result should be divisible by the generator polynomial to give zero remainder.

$$
\begin{array}{r}
6x^2 + 7x + 4 \\
x^4 + 4x^3 + 7x^2 + 7x + 5 \overline{)\, 6x^6 + 2x^5 + 1x^4 + 1x^3 + 1x^2 + 7x + 2} \\
\underline{6x^6 + 5x^5 + 4x^4 + 4x^3 + 3x^2} \\
7x^5 + 5x^4 + 5x^3 + 2x^2 + 7x \\
\underline{7x^5 + 1x^4 + 3x^3 + 3x^2 + 6x} \\
4x^4 + 6x^3 + 1x^2 + 1x + 2 \\
\underline{4x^4 + 6x^3 + 1x^2 + 1x + 2} \\
0x^4 + 0x^3 + 0x^2 + 0x + 0
\end{array}
$$

As the remainder is zero, the RS code is correct. Now for the decoding and error correction. Let an error be introduced such that the $1x^4$ term becomes $7x^4$. We proceed as before by dividing the received sequence by the generator polynomial to give, in this case, a remainder. Rather than use a polynomial with the powers of x, we will simply use the magnitudes, i.e. $x^4 + 4x^3 + 7x^2 + 7x + 5$ becomes 14775 and $6x^6 + 2x^5 + 7x^4 + 1x^3 + 1x^2 + 7x + 2$ becomes 6271172. Thus

$$
\begin{array}{r}
672 \\
14775\overline{)6271172} \\
\underline{65443} \\
73527 \\
\underline{71336} \\
26112 \\
\underline{23551} \\
5443
\end{array}
$$

This remainder of 5443 ($5x^3 + 4x^2 + 4x + 3$) shows that there is an error in the received block of data with the remainder being unique to the magnitude and location of the error. So, we can generate a look-up table of remainders and use it to correct the error. Table 3.8 shows the remainders and their location.

To return to our example, division of the received word 6271172 by the generating polynomial 14775 gives us a remainder of 5443 (highlighted in Table 3.8). This tells us that an error has occurred on the x^4 term. To correct for this error, we need to change the coefficient to 1 (remainder 0000) by adding 6. (Recall that subtraction is the same as addition.) A look-up table is very attractive, simple and intuitive. However, it is very resource intensive particularly if the code is large. There is an alternative based on syndromes and some rather complex mathematics. As this is an introductory text, it is not considered here.

TABLE 3.8
Look-up Table for Generated Remainders

	1	2	3	4	5	6	7
x^6	5266	3722	1477	4611	6544	0000	2355
x^5	3176	0000	1642	6257	7415	4563	5156
x^4	0000	7224	3551	2667	5212	4336	**5443**
x^3	0000	3000	2000	5000	4000	7000	6000
x^2	0000	0300	0200	0500	0400	0700	0600
x^1	0060	0050	0040	0030	0020	0010	0000
x^0	0003	0000	0001	0002	0007	0004	0005

3.5 PROBLEMS

1. An ADC has a range of 0–12 V and uses 14 bits of code. The sampling frequency is 100 kHz. Determine
   a. The quantisation noise.
   b. The maximum signal-to-quantisation noise ratio.
   c. The minimum signal voltage given that the minimum signal-to-quantisation noise ratio is 30 dB.
   d. The serial data rate out of the ADC.
   [211 µV; 40214 or 212 dB; 6.7 mV rms; 1.4 Mbit/s]

2. In a certain digital link, a logic one is represented by 0.5 V and a logic zero by 0 V. If the noise is 100 mV rms, determine the probability of error.
   [2.8×10^{-7}]

3. For the BPSK demodulator using a matched filter, derive an expression for the bit error rate in terms of the pulse energy and the noise spectral density. If E/η is 3.5, determine the error rate.
   [372 errors in 10^9]

4. Repeat Q3 for 4-QAM.
   [6.57 errors in 10^3]

5. With reference to Table 3.5, determine the maximum likelihood codeword if the detected codeword is 11 00 01 10 00 00 11.

6. With reference to Table 3.6, plot the Viterbi path for the codeword 00 00 01 11 10 01 00.

4 Introduction to Analogue Modulation

In the previous chapter, we concentrated on digital modulation because all TV is now digital and we also have digital radio. However, analogue radio systems are still in existence and that is something we are concerned with here.

Consider a high-frequency carrier given by

$$v_c(t) = V_c \cos(\omega_c t + \varphi_c) \tag{4.1}$$

where:

subscript c indicates it is the carrier that is being referred to

V_c is the carrier amplitude (peak)

ω_c is the angular frequency of the carrier

φ_c is the phase offset from zero

Although we have used the cosine representation, remember that a cosine is simply a sine shifted by 90°.

There are three parameters that can be altered in Equation 4.1: the amplitude (giving amplitude modulation or AM), the frequency (giving frequency modulation or FM) and the phase (giving phase modulation or PM).

4.1 AMPLITUDE MODULATION

This is a very popular form of modulation for broadcasting – AM stations can be found on the long-wave, medium-wave and short-wave bands. As the name suggests, it is the amplitude of the carrier signal, V_c, that is changed according to the modulation, which could be audio or even data. So, instead of having simply V_c, we have a carrier amplitude of $V_c + v_m(t)$ where

$$v_m(t) = V_m \cos \omega_m t \tag{4.2}$$

It is useful to put some frequencies in at this time to remind us that we are dealing with audio (AF) and radio frequencies (RF). Let us choose 1 kHz for the AF and 1 MHz for the RF. With reference to Figure 4.1, the RF signal has a complete revolution every 1 μs whereas the AF signal rotates every 1 ms. This is an important point and the reason why will become clear very shortly.

We are amplitude modulating the carrier given in Equation 4.1 and we ignore the phase for the moment. So, the carrier amplitude becomes

$$V_c + v_m(t) = V_c + V_m \cos \omega_m t \tag{4.3}$$

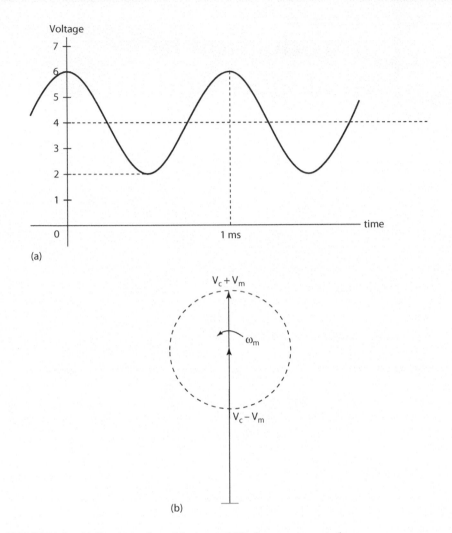

(a)

(b)

FIGURE 4.1   (a) Carrier and modulation and (b) phasor representation.

which is simply a cosine wave with a direct current (dc) level of V_c on it. Taking V_c of 4 V and V_m of 2 V results in the plot of Figure 4.1a. The rotating phasor diagram that produces the graph of Figure 4.1a is shown in Figure 4.1b.

The audio signal has been added to the amplitude of the carrier and so the resultant signal is

$$v_{AM}(t) = (V_c + V_m \cos \omega_m t) \cos \omega_c t \tag{4.4}$$

$$= V_c \cos \omega_c t + V_m \cos \omega_m t . \cos \omega_c t \tag{4.5}$$

The second term in Equation 4.5 is the product of two cosines. Thus,

$$v_{AM}(t) = V_c \cos\omega_c t + \frac{V_m}{2}\{\cos(\omega_c + \omega_m)t + \cos(\omega_c - \omega_m)t\} \tag{4.6}$$

There are a number of points to note about Equation 4.6. The first term is the original carrier – both amplitude and frequency. (We will return to this presently.) The second term has the amplitude of the modulation. (It is actually half the amplitude but there are two components in the following bracket.) In the bracket are two terms: one at a frequency higher than the carrier by the modulation frequency and one at a frequency lower than the carrier by the modulation frequency. The first term is known as the upper side frequency (USF) and the second term is the lower side frequency (LSF). Taking $f_c = 1$ MHz and $f_m = 1$ kHz (as before), we get a carrier frequency of 1 MHz, a USF of 1.001 MHz and an LSF of 0.999 kHz. This is shown in Figure 4.2. As noted in the Introduction, individual stations can transmit on different frequencies. Thus, we could have two stations at carrier frequencies of 1 and 1.2 MHz. In reality, the channel spacing for an AM broadcast is 10 kHz; thus, the bandwidth of the AM signal is 10 kHz, giving a maximum modulating frequency of 5 kHz.

If the amplitude of the modulating signal increases, the amplitude of the side frequencies will also change. If the frequency of the modulation increases, the frequency of the upper component will increase but that of the lower component will decrease. We can see this if we take a trapezoidal representation of the modulating signal as in Figure 4.3.

Figure 4.3a shows the baseband with a maximum frequency of f_m and Figure 4.3b shows the carrier signal as before. Figure 4.3c shows the resultant spectrum

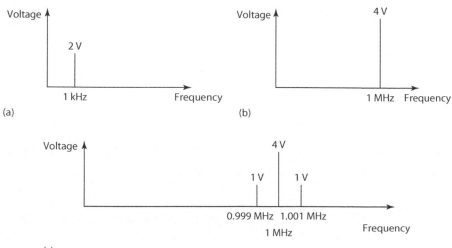

FIGURE 4.2 Spectra of (a) baseband, (b) carrier and (c) AM signal.

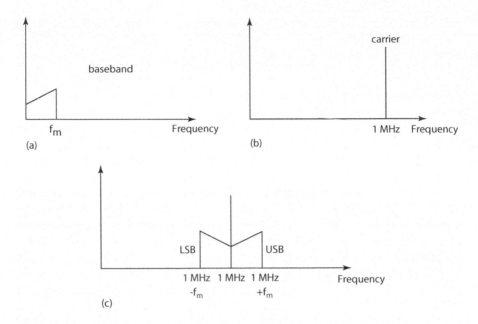

FIGURE 4.3 Baseband modulation of a carrier.

consisting of the upper sideband (USB) and the lower sideband (LSB). So, by modulating the amplitude of the carrier, the baseband signal appears either side of the carrier frequency. Thus, we have achieved what we wanted to do – the audible signal has been converted into a high-frequency electromagnetic wave and other stations can use different frequencies. This is a very important reason why we modulate a high-frequency carrier.

The time response of an AM signal is shown in Figure 4.4a. Note that in this figure, the frequency of the carrier is only 10 times that of the modulating signal. This has been deliberately chosen so that the carrier can be seen. The phasor diagram that gives rise to the time response is shown in Figure 4.4b. This phasor diagram is useful when we study the noise performance of AM. A simple AM modulator can be made using a voltage-controlled amplifier (VCA). Such a device can be based around a dual gate *metal-oxide semiconductor field-effect transistor* (MOSFET) (Figure 4.5). The carrier signal is amplified by the MOSFET and the tuned circuit is the frequency-selective amplifier load. The audio signal changes the gain of the amplifier and so the carrier amplitude is increased and decreased according to the audio amplitude.

An important parameter when discussing the performance of AM is the modulation depth, m. This is a measure of how much of the carrier is used by the modulation. It is defined as the ratio of the modulating voltage to the carrier voltage. Thus,

$$m = \frac{V_m}{V_c}$$

(4.7)

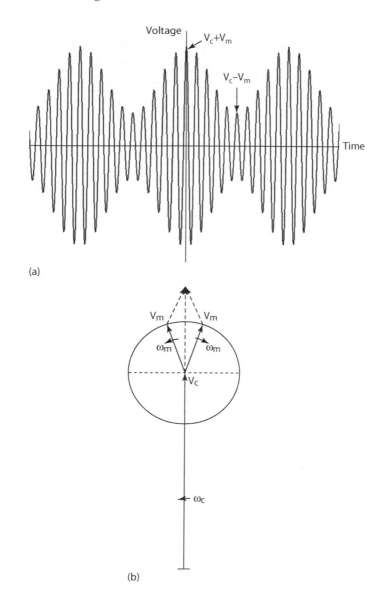

(a)

(b)

FIGURE 4.4 (a) Time response of an AM signal; (b) phasor diagram for AM.

The modulation depth can vary from 0 (no modulation) to beyond 100% (over-modulation). In most communications links using AM, m is very close to 100%. This is because it is more power efficient (not normally a problem for broadcasters) and it offers a better signal to noise ratio (S/N; of importance to all users). Figure 4.6 shows the AM signal resulting from $m = 10\%$, 100% and 120%. Here, the modulating frequency is 30 times lower than the carrier. This is to make the difference between the m values clear. It should be noted that the plot for $m = 120\%$ shows distortion. This will translate into distortion of the audio signal when it is demodulated and also

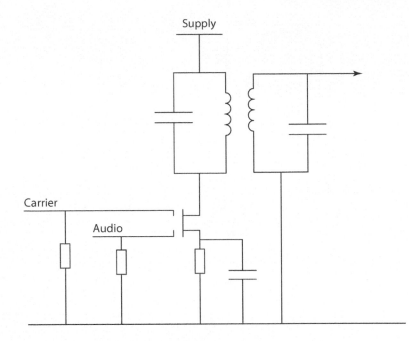

FIGURE 4.5   A simple dual gate MOSFET modulator.

generate interference to other users. Of course, it is impossible to guarantee that the audio signal will never cause overmodulation; however, an audio circuit known as a vogad – voice-operated gain adjusting device – can be used to reduce the range of amplitudes presented to the modulator.

Mention has already been made of the power efficiency of AM. There are three components that make up an AM signal and each of these contributes to the total power. The carrier power is

$$P_c = \frac{\left(\dfrac{V_c}{\sqrt{2}}\right)^2}{R} = \frac{V_c^2}{2R} \tag{4.8}$$

The presence of the $\sqrt{2}$ under the V_c^2 term is due to the use of the root mean square (rms) voltage to give the average power. The two sidebands have the same amplitude ($V_m/2$) and so the sideband power in each is

$$P_{side} = \frac{\left(\dfrac{V_m}{2\sqrt{2}}\right)^2}{R} = \frac{V_m^2}{8R} = \frac{mV_c^2}{8R} \tag{4.9}$$

where we have used the fact that $m = V_m/V_c$. So, the ratio of the carrier power to the total power is

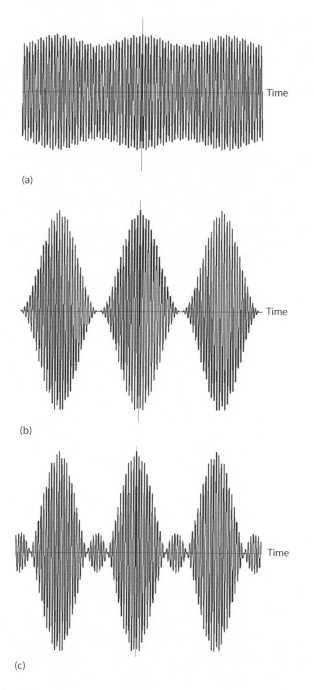

(a)

(b)

(c)

FIGURE 4.6 An AM signal with different modulation depths: (a) $m = 10\%$; (b) $m = 100\%$; (c) $m = 120\%$.

$$\frac{P_c}{P_c + 2P_{\text{side}}} = \frac{1}{1 + 2P_{\text{side}} / P_c}$$

$$= \frac{1}{1 + \dfrac{2\dfrac{m^2 V_c^2}{8R}}{\dfrac{V_c^2}{2R}}} \qquad\qquad (4.10)$$

$$= \frac{1}{1 + m^2 / 2}$$

If the carrier is completely modulated, $m = 1$ and 66.6% of the total power is in the carrier. This is a significant result. We have just calculated the amount of total power in a component whose amplitude and frequency do not change with the audio. In other words, it is wasted power – 66.6% of the power in the signal is wasted. This is not so significant if the station is a broadcaster using power from the grid. However, if AM is used in battery-powered portable transmitters, the waste of power is a problem. Portable transmitters such as walkie-talkies and mobile phones do not use AM as power efficiency is a major issue.

We have seen that the process of modulation results in the baseband being attached by some means to a high-frequency carrier. The process of demodulation is to get the baseband signal back again. For AM demodulation, a very simple circuit using a diode, capacitor and resistor can be used. This is called the *diode demodulator* (Figure 4.7).

The diode in Figure 4.7 is configured as a half-wave rectifier which removes the negative going half cycle of the AM signal (Figure 4.8). The resistor-capacitor (RC) filter acts to remove the IF component leaving the envelope behind. The advantage of the diode demodulator is that it is inexpensive requiring easily sourced components. There are disadvantages, however. The diode requires 0.2 V to turn on which impacts on the modulation depth; the diode introduces distortion; and the cut-off frequency of the filter is critical. The cut-off frequency of the RC filter must be set to remove the intermediate frequency (IF) component and leave the audio. Taking an IF of 470 kHz and a maximum AF of 20 kHz (for music), the filter must have a

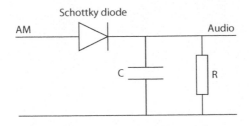

FIGURE 4.7   A simple diode demodulator.

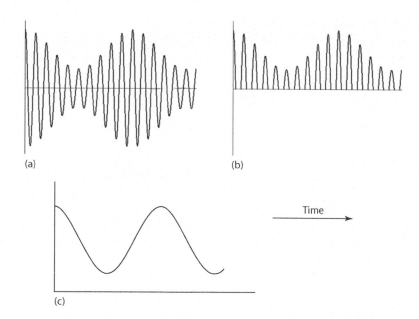

FIGURE 4.8 Waveforms showing demodulation by diode (a) input to demodulator; (b) half-wave rectification; and (c) demodulated waveform.

cut-off frequency of 20 kHz. As the IF is more than a decade above this cut-off, it will be sufficiently attenuated.

The distortion in this demodulator comes from the diode characteristic itself. The current, I_D, through a forward-biased diode approximates to

$$I_D = I_o \exp\left(\frac{V}{\eta V_T}\right) \tag{4.11}$$

where:

I_o is the reverse leakage current
V is the forward-bias voltage across the diode
η is the emission coefficient
V_T is the thermal voltage (25 mV at room temperature)

This exponential form can be approximated to an infinite series such as

$$I_D = I_o\left(a + bx + cx^2 + dx^3 + \text{etc.}\right) \tag{4.12}$$

where x is $V/\eta V_T$. We are interested in the demodulation of AM and so we can take the voltage as the AM signal. Thus,

$$x = \left(V_c + V_m \cos \omega_m t\right)\cos \omega_{IF} t.\text{constant} \tag{4.13}$$

In a moment, we will take the ratio of two amplitudes and so we will ignore the constant in Equation 4.13 as it will cancel out anyway. There are four components of interest in Equation 4.12. The first is simply a dc term while the second is the original AM signal. Neither of these is wanted and can be ignored. The expansion of the third term yields

$$\left(V_c + V_m \cos\omega_m t\right)^2 \cos^2 \omega_{IF} t$$

$$= \left(V_c^2 + 2V_c V_m \cos\omega_m t + V_m^2 \cos^2 \omega_m t\right) \frac{1}{2}\left(1 + \cos 2\omega_{IF} t\right) \qquad (4.14)$$

$$= \frac{1}{2}\left(V_c^2 + 2V_c V_m \cos\omega_m t + \frac{V_m^2}{2}\left(1 + \cos 2\omega_m t\right)\right)\left(1 + \cos 2\omega_{IF} t\right)$$

Rather than continue to expand Equation 4.14 further, it is instructive to examine the frequency components. The final bracket multiplies the first bracket by 1 or $\cos 2\omega_{IF} t$. Recall that multiplying two cosines together gives sum and difference components so we expect to see components at $2\omega_{IF}$. As we are using a low-pass filter (LPF), components at $2\omega_{IF}$ can be ignored. Consequently, we have three frequency components: dc, ω_m and $2\omega_m$. The dc component can easily be taken care of by using a capacitor. The ω_m term is the frequency we require – it is the baseband. Unfortunately, there is the third component at $2\omega_m$ and this is a problem. A baseband can cover 20 Hz–5 kHz and so there is a mixing product at 40 Hz to 10 kHz. This represents interference. The ratio of unwanted to wanted signals is

$$\frac{V_m^2/2}{2V_c V_m} = \frac{m^2 V_c^2}{4m V_c^2} \qquad (4.15)$$

$$= \frac{m}{4}$$

Note that in taking the ratio of two components, the scaling factors that we ignored are cancelled out. Equation 4.15 shows that undesirable intermodulation products are produced and cannot be removed by filtering. In addition, their amplitude is ¼ that of the demodulated signal. As they come from exactly the diode component we need to demodulate, there is little we can do except try a different demodulator.

In the spectrum of AM, the baseband signal appears either side of the carrier. In the examples used, the sideband signals are at 1 MHz–1 kHz and 1 MHz + 1 kHz. These signals come from the product of the carrier and the audio. With the demodulator we are about to consider, the AM is mixed with a local carrier signal thereby producing the baseband directly.

The principle of operation is very simple. The AM signal at the IF is mixed with the output of a local oscillator operating at the IF (Figure 4.9). The resultant is an AM at twice the IF (which can be filtered out) and the baseband signal:

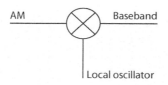

FIGURE 4.9 Coherent demodulation.

$$\text{Mixer output} = \left(V_c + V_m \cos\omega_m t\right)\cos\omega_{\text{IF}} t . V_{\text{LO}} \cos\omega_{\text{IF}} t$$

$$= V_{\text{LO}}.\left(V_c + V_m \cos\omega_m t\right).\frac{1}{2}\left(1 + \cos 2\omega_{\text{IF}} t\right) \qquad (4.16)$$

$$= \frac{V_{\text{LO}}V_c}{2} + \frac{V_{\text{LO}}V_m}{2}\cos\omega_m t + 2\omega_{\text{IF}} \text{ terms}$$

It is clear from Equation 4.16 that the baseband signal has been recovered and that it has had its amplitude increased by the amplitude of the local oscillator. Note also that there is a useful dc term that depends on the received carrier voltage. This can be used in an automatic gain control system, which helps when you are listening to the radio in the car and the signal fades due to obstructions. Carrier recovery can be achieved by using a phase-lock loop (PLL). There are problems associated with mixing down to the baseband: the carrier frequency and phase must be correct.

$$\text{Mixer output} = \left(V_c + V_m \cos\omega_m t\right)\cos\omega_{\text{IF}} t . V_{\text{LO}} \cos\left(\omega_{\text{IF}} + \Delta\omega\right)t$$

$$= V_{\text{LO}}.\left(V_c + V_m \cos\omega_m t\right).\frac{1}{2}\left(\cos\Delta\omega t + \cos\left(2\omega_{\text{IF}} + \Delta\omega\right)t\right) \quad (4.17)$$

$$= \frac{V_{\text{LO}}V_c}{2}\cos\Delta\omega t + \frac{V_{\text{LO}}V_m}{2}\cos\omega_m t \cos\Delta\omega t + 2\omega_{\text{IF}} \text{ terms}$$

An examination of Equation 4.17 reveals that the first term is a carrier at a frequency of $\Delta\omega$ while the second term generates two sidebands either side of $\Delta\omega$. This is shown diagrammatically in Figure 4.10. It should be remembered that the cosine of a negative number is the same as the cosine of a positive number. So, taking the ideal situation, the LSB folds over around zero. If there is a frequency offset, as in Figure 4.10b, the fold over occurs at zero as before but symmetry is lost and distortion results. The effect of phase offsets is left to the reader.

Figure 4.11 shows the phasor diagram which has the modulation and the in-phase and quadrature noise terms. Note that because we are considering the carrier as a sine wave, the sine term of the noise is the in-phase term and the cosine term is the quadrature term. The situation would be reversed if we used the cosine representation of AM. As can be seen from Figure 4.11, the resultant AM signal is made up of the phasor sum of the noiseless AM signal and the in-phase and quadrature components. If we take the large signal condition, $V_c \gg$ noise, the resultant will simply be the sum of the carrier, the modulation and the in-phase noise. (This is a reasonable assumption because the S/N can be quite large.) Thus,

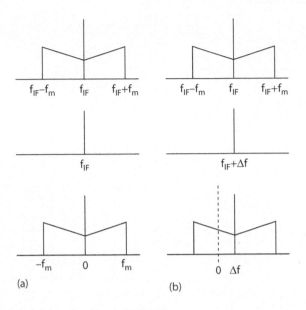

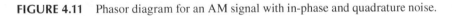

(a)   (b)

FIGURE 4.10 Distortion caused by frequency offsets (a) correct demodulation and (b) frequency offset distortion.

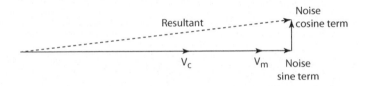

FIGURE 4.11 Phasor diagram for an AM signal with in-phase and quadrature noise.

$$\text{AM} + \text{noise} = \left(V_c + V_m \sin \omega_m t + n(t) \right) \sin \omega_{\text{IF}} t \tag{4.18}$$

where

$$n(t) = \sqrt{2\eta df} \, \sin \omega_{\text{off}} t \tag{4.19}$$

We want to find the S/N for AM. As shown in Figure 4.11, the noise adds to the AM signal and therefore the envelope is affected by noise. Assuming an ideal envelope detector, the recovered modulation will be

$$\text{Modulation} = V_m \sin \omega_m t + n(t)$$

(We could consider the non-ideal diode demodulator. This is left to the interested reader.) The peak amplitude of the audio signal is V_m and so the audio power is

$$\left(\frac{V_m}{\sqrt{2}}\right)^2 \Big/ R = \frac{m^2 V_c^2}{2} \Big/ R$$

The noise in a strip at ω_{off} within the IF is given by Equation 2.17 and so the noise power in the strip is

$$\left(\frac{\sqrt{2\eta df}}{\sqrt{2}}\right)^2 \Big/ R = \eta df \Big/ R$$

To find the total noise on the AM signal, we need to integrate this expression with respect to the frequency over the bandwidth of the IF (B Hz). So, the total noise power is

$$\int_{-B/2}^{B/2} \eta df \Big/ R = \eta B \Big/ R$$

Thus, the audio S/N is

$$\frac{S}{N} = \frac{\dfrac{m^2 V_c^2}{2} \Big/ R}{\eta B \Big/ R} = m^2 \frac{C}{N} \qquad (4.20)$$

where:
  m is the modulation depth
  C/N is the carrier to noise ratio at the input to the detector

As previously discussed, it is advantageous to operate with $m = 100\%$ as the power efficiency is highest. This condition also gives the best audio S/N and that is very important. A target S/N is 60 dB or 10^6. This should be compared to the S/N parameter for digital signals.

It is interesting to calculate the S/N for coherent detection. As before, we have in-phase and quadrature noise and, if the carrier is large enough, the effects of the quadrature noise can be ignored. Thus, the signal plus noise applied to the coherent detector (mixer) is

$$AM + noise = \left(V_c + V_m \sin \omega_m t + n(t)\right) \sin \omega_{IF} t \qquad (4.21)$$

Thus, the output of the mixer is

$$\left(V_c + V_m \sin \omega_m t + n(t)\right) \sin^2 \omega_{IF} t$$

which, after filtering, gives $\frac{1}{2}\left(V_c + V_m \sin \omega_m t + n(t)\right)$.

The first term in the bracket is dc and so it can be removed by passing through a capacitor. The second term is the modulation with a power of $\left(V_m / \sqrt{2}\right)^2 / R$ and the last term is the noise with a power of $\eta B/R$. Thus, the S/N is the same as the ideal envelope detector, i.e.

$$\frac{S}{N} = m^2 \frac{C}{N} \tag{4.22}$$

4.2   DOUBLE SIDEBAND SUPPRESSED CARRIER MODULATION

In the previous section, we examined the modulation technique known as AM. We saw that the power efficiency was quite low with 66.6% of the total power wasted in the carrier. With the modulation format known as *double sideband suppressed carrier* (DSB-SC), the carrier component is not produced at all. So, the spectrum of DSB-SC only contains the two sidebands (Figure 4.12).

The production of DSB-SC is mathematically very simple; we just multiply the audio and the carrier using a mixer (Figure 4.13). Thus,

$$v_{\text{DSB-SC}}(t) = V_m \cos \omega_m t . V_c \cos \omega_{\text{IF}} t$$
$$= \tfrac{1}{2} V_m V_c \left\{ \cos\left(\omega_{\text{IF}} + \omega_m\right)t + \cos\left(\omega_{\text{IF}} - \omega_m\right)t \right\} \tag{4.23}$$

Equation 4.23 shows that we only have the two sidebands and no carrier. DSB-SC is very efficient as there is no wasted power in the carrier. However, the power efficiency is not 100% as there are two sidebands both carrying the same information. Thus, we could say that the power efficiency is 50%.

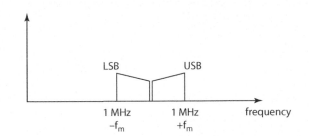

FIGURE 4.12   Spectrum of DSB-SC.

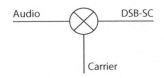

FIGURE 4.13   Mixer generation of DSB-SC.

Demodulation is carried out by mixing down to the baseband, but it suffers from the same problem if the frequency is not exactly at the IF. The Costas loop of Figure 3.18 can be used to extract the missing carrier.

The noise performance can be found by

$$\text{DSB-SC} = V_{\text{DSB}} \sin \omega_m t \, \sin \omega_{\text{IF}} t + x(t) \sin \omega_{\text{IF}} t + y(t) \cos \omega_{\text{IF}} t$$

Therefore, the baseband is (after filtering)

$$\frac{V_{\text{DSB}}}{2} \sin \omega_m t + \frac{x(t)}{2} + 0$$

Note that the quadrature term mixes to zero. Thus, the S/N is (cancelling the factor of 2):

$$\frac{S}{N} = \frac{\left(\dfrac{V_{\text{DSB}}}{\sqrt{2}}\right)^2 \Big/ R}{\dfrac{\eta B}{R}} \tag{4.24}$$

$$= \frac{(V_{\text{DSB}})^2}{2\eta B}$$

4.3 SINGLE SIDEBAND MODULATION

Instead of transmitting two sidebands as in DSB-SC, a filter can be used to remove one or the other sideband (Figure 4.14). The result is USB or LSB. There is something rather special about the sideband filter. Consider the DSB-SC signal and, in particular, the spectral components around the carrier frequency (Figure 4.15). The sideband filter needs to filter out either the LSB or the USB. As Figure 4.15 shows, the separation of the two sidebands is 600 Hz if it is assumed that the audio signal is limited to 300 Hz–3.4 kHz. A tuned circuit will not be able to adequately filter out the unwanted sideband, neither will an RC network. What is needed is a special filter that has a very sharp cut-off. Such a filter is a crystal filter based around quartz crystals. A crystal in radio terms is a piece of quartz crystal that has been cut to

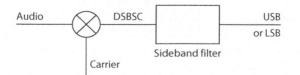

FIGURE 4.14 Generation of an SSB using DSB-SC and a filter.

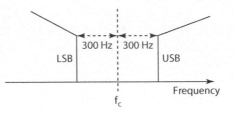

FIGURE 4.15 Illustrating the need for precision sideband filters.

FIGURE 4.16 Circuit symbol of a crystal.

resonate at a particular frequency. It is a physical resonance that also translates into an electrical one – the piezoelectric effect. The crystal is placed between the plates of a capacitor (Figure 4.16). So, an electric field at the correct frequency causes an oscillation. Crystals can be used in sideband filters but, in the interest of production, sideband filters are designed to operate at a standard frequency such as 9 MHz. As this is unlikely to be the desired transmit frequency, mixers are used to change frequency. Crystals are also used in oscillators where they provide a very accurate frequency source – watches use crystal oscillators.

Consider a single sideband (SSB) modulator operating at a frequency of 9 MHz. The final frequency of transmission is to be 100 MHz. A possible solution is shown in Figure 4.17. We require a total frequency shift of 91 MHz. This can be achieved in one shift but there is a problem with such a large change in frequency. Coming out of the mixer are sum and difference frequencies. So, mixing 9 MHz with 91 MHz will yield 100 MHz and 82 MHz. These are only 18 MHz apart and it will be difficult to separate the two frequencies using a tuned circuit because the Q will be small at this frequency (Appendix III). So, the solution adopted in Figure 4.17 is to shift by a relatively small amount initially to generate 29 MHz with a split in frequency of 18 MHz. This is well within the capabilities of a 29 MHz tuned circuit. The final shift is from 29 to 100 MHz with a split in frequency of 58 MHz. Again, this is within the capability of a high-frequency 100 MHz tuned circuit. The power amplifier in a transmitter is usually linear so that the signal is not distorted.

In order to demodulate an SSB, the baseband is recovered by mixing with the IF. There are, however, some problems. If the local oscillator is not at exactly the right frequency, distortion occurs. However, the situation is not as problematic as with DSB-SC because there is only one sideband to consider:

$$\text{Mixer output} = V_{\text{SSB}} \cos\left(\omega_{\text{IF}} + \omega_m\right)t . \cos\left(\omega_{\text{IF}} + \Delta\omega\right)t$$

$$= \frac{V_{\text{SSB}}}{2} . \left(\cos\left(\omega_m - \Delta\omega\right)t + \cos\left(2\omega_{\text{IF}} + \omega_m + \Delta\omega\right)t\right) \tag{4.25}$$

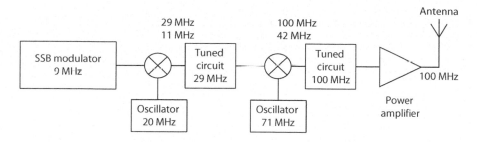

FIGURE 4.17 Possible implementation of an SSB transmitter.

As shown in Equation 4.25, a positive frequency offset in the local oscillator distorts the baseband by shifting it by $\Delta\omega$. When this occurs, the radio operator will retune so that the offset is negative. This will move the baseband up and then the person speaking via the radio will sound a little like Donald Duck. This is an unfortunate by-product of having a modulation format that is 100% power efficient and occupies a bandwidth equal to the audio signal. Phase also presents a problem and readers are referred to the Problems section for this chapter.

Taking a local oscillator with the correct frequency, the mixer output will be

$$\text{Mixer output} = \left(V_{\text{SSB}}\cos\left(\omega_{\text{IF}}+\omega_m\right)t+x(t)\cos\omega_{\text{IF}}t+y(t)\sin\omega_{\text{IF}}t\right).\cos\omega_{\text{IF}}t$$

$$= \frac{V_{\text{SSB}}}{2}.\cos\omega_m t+\frac{x(t)}{2} \tag{4.26}$$

Cancelling the factor of half and the resistance, the S/N is

$$\frac{S}{N}=\frac{\left(\dfrac{V_{\text{SSB}}}{\sqrt{2}}\right)^2}{\eta B}=\frac{V_{\text{SSB}}^{\;2}}{2\eta B} \tag{4.27}$$

We can determine a figure of merit as the S/N after the detector versus the S/N before the detector. This exercise appears in the problems for this chapter.

4.4 FREQUENCY MODULATION

In the form of modulation known as *frequency modulation*, it is the frequency of the carrier that varies, not the amplitude. This has a number of benefits as we will see. First of all, consider a cosine signal given by

$$v(t)=\cos\theta(t)$$

$\theta(t)$ is the phase in radians. We can generate our more familiar form as follows:

$$\theta(t) = \int 2\pi f \, dt$$

(4.28)

$$= \omega t$$

Hence, our carrier becomes

$$v(t) = \cos \omega t$$

We don't have constant frequency with FM because it is generated by applying the modulation to a voltage-controlled oscillator (VCO). If the modulation increases in amplitude, the frequency of the VCO increases. A decrease in modulation amplitude reduces the VCO frequency. There is a constant of proportionality which we will call k_f with units of Hertz per volt (Hz/V). So, the modulated carrier frequency will be given by

$$f = f_c + k_f . V_m \cos \omega_m t$$

(4.29)

where we have used our usual form for the modulation. This has units of Hertz and, in order to convert into an angle, we need to multiply by 2π and integrate with respect to time. Thus,

$$\theta_{\mathrm{FM}}(t) = 2\pi \int \left(f_c + k_f . V_m \cos \omega_m t \right) dt$$

$$= \left(2\pi f_c + 2\pi \frac{k_f V_m}{\omega_m} \sin \omega_m t \right) t$$

$$= \left(\omega_c + \frac{k_f V_m}{f_m} \sin \omega_m t \right) t$$

The resultant FM signal is then given by

$$v_{\mathrm{FM}}(t) = V_c \cos \left(\omega_c + \frac{k_f V_m}{f_m} \sin \omega_m t \right) t$$

(4.30)

The first term in the bracket in Equation 4.30 is simply the original carrier frequency. The second term introduces a phase shift. The term $k_f V_m$ is called the *maximum frequency deviation* and has the symbol Δf. The dimensionless modulation index (not to be confused with modulation depth) is $\Delta f / f_m$. So,

$$\Delta f = k_f V_m$$

(4.31)

$$m_f = \frac{k_f V_m}{f_m} = \frac{\Delta f}{f_m}$$

(4.32)

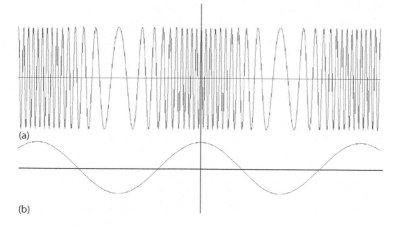

(b)

FIGURE 4.18 (a) An FM signal and (b) its modulation.

Hence,

$$v_{FM}(t) = V_c \cos(\omega_c + m_f \sin \omega_m t)t \qquad (4.33)$$

Figure 4.18 shows the time plot of Equation 4.33. It is clear that the frequency changes according to the audio (the sine wave).

Figure 4.18a shows an FM signal. As can be seen, the frequency varies according to the modulation shown in Figure 4.18b. A point worth noting is that the amplitude of the modulated carrier is fixed. This is totally different to AM and leads to FM having a power efficiency of 100%, although this does come at the cost of a large bandwidth.

We can obtain the spectrum of FM by expanding Equation 4.33. Unfortunately, this is not as easy as AM:

$$
\begin{aligned}
v_{FM}(t) &= V_c \cos(\omega_c + m_f \sin \omega_m t)t \\
&= V_c \{\cos \omega_c t \cos(m_f \sin \omega_m t)t - \sin \omega_c t \sin(m_f \sin \omega_m t)t\}
\end{aligned}
\qquad (4.34)
$$

In Equation 4.34, we have two terms that cause a problem – the cosine of a sine and the sine of a sine. There is a solution using Bessel functions of the first kind. Appendix X shows how the coefficients vary with the modulation index and Table AX.1 details the coefficients. The expansion of Equation 4.34 is

$$
\begin{aligned}
v_{FM}(t) = {}& J_0(m_f) V_c \{\cos \omega_c t\} \\
&+ J_1(m_f) V_c \{\cos(\omega_c + \omega_m)t - \cos(\omega_c - \omega_m)t\} \\
&+ J_2(m_f) V_c \{\cos(\omega_c + 2\omega_m)t + \cos(\omega_c - 2\omega_m)t\} \\
&+ J_3(m_f) V_c \{\cos(\omega_c + 3\omega_m)t - \cos(\omega_c - 3\omega_m)t\} + \text{etc.}
\end{aligned}
\qquad (4.35)
$$

The first term is the carrier and it should be noted that it varies according to m_f which is directly proportional to the audio amplitude. This is totally different to AM in which the carrier amplitude remained constant. The second point to note is that the LSBs alternate in sign – if $J_n(m_f)$ is odd, the LSB is negative (this is simply a phase shift). The next thing to note is the number of side frequencies – we have an infinite number of harmonics of the audio signal. We can ignore amplitudes that are less than 10% of V_c and so the spectrum of FM with an m_f of 3 is as drawn in Figure 4.19. The Bessel function coefficients carry on to infinity and so it might be thought that the bandwidth of FM is also infinite. In theory, this is true but, in practice, we can ignore coefficients that are less than 10%. This is known as *Carson's rule* and the bandwidth of the FM signal is

$$\text{Bandwidth} = 2\left(m_f + 1\right) f_m \qquad (4.36)$$

Examination of the table of Bessel coefficients for an m_f of 3 reveals four sets of side frequencies and so the bandwidth is $8f_m$. This is termed *wideband FM* (WBFM). The power efficiency of FM is 100% as can be seen by summing the square of the Bessel coefficients for a particular value of m_f. (Note that the table in Appendix X is not completely accurate due to insufficient space for all the coefficients.) WBFM is transmitted on very high frequency (VHF) and above as the bandwidth required for transmission is large. (If WBFM were transmitted on lower frequencies, it would take up the whole of LW and MW and most of SW.)

There is a very special case of FM and that is when $m_f < 0.28$. With this m_f value, there are only two significant sidebands – one upper and one lower. This is called *narrowband FM* (NBFM), and it is very important because the bandwidth is $2f_m$, like AM, but it is 100% power efficient. Thus, we do not have to carry heavy batteries around when using portable transmitting equipment.

The generation of FM can be done by various means: a variable capacitance (varicap) diode in a tuned circuit, a specialist VCO such as the NE566, a PLL or the Armstrong method. For the first, a numeric example will help. Figure 4.20 shows a

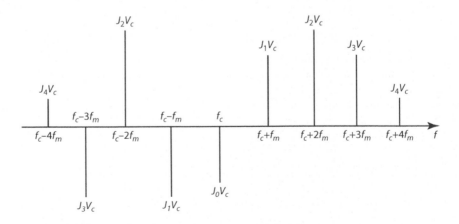

FIGURE 4.19   The FM spectrum with $m_f = 3$.

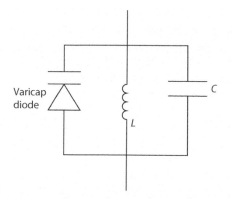

FIGURE 4.20   A varicap diode–controlled tuned circuit.

parallel tuned circuit, using a varicap diode, as part of an oscillator circuit. We will assume the following:

$$100 \text{ pF} \le C_{\text{diode}} \le 300 \text{ pF} \quad C = 10 \text{ nF} \quad L = 6.21 \text{ mH}$$

The inductor value is very precise but, in practice, a movable ferrite core can be used in the inductor to fine-tune the frequency of the oscillator. Using these values, the nominal resonant frequency is

$$f_{\text{nom}} = \frac{1}{2\pi\sqrt{\left(C + C_{\text{diode nom}}\right)L}}$$

$$= \frac{1}{2\pi\sqrt{\left(10 \text{ nF} + 200 \text{ pF}\right)6.21 \text{ mH}}}$$

$$= 20 \text{ kHz}$$

The maximum frequency of oscillation is when C_{diode} is 100 pF. By following the same procedure, this maximum is 20.096 kHz. This change in capacitance comes about from the minimum audio amplitude and so the maximum frequency deviation, Δf, is 96 Hz. Getting a larger deviation is difficult because of non-linearities in the derivation (the root relationship between frequency and capacitance). The maximum frequency deviation for broadcast purposes is 75 kHz. We can achieve this, or get close, by passing the signal through harmonic generators and frequency shifters.

Figure 4.21 shows how this is achieved using a system of harmonic generators and frequency shifters to get our 20 kHz carrier with 96 Hz deviation into an FM signal with Δf close to 75 kHz and a centre frequency of 100 MHz. Whenever the signal passes through a harmonic generator, the frequency deviation increases by the harmonic order. Consider passing the 20 kHz, 96 Hz signal through a second harmonic generator. The carrier will become 40 kHz, but what of the deviation? Recall the definition of Δf – it is by how much the carrier moves when a dc signal of amplitude equal to the maximum modulation is applied to the VCO. So, our carrier is moved by 96 Hz to give a signal at 20.096 kHz. If this is applied to the harmonic generator, the new frequency will be 40.192 kHz and so the Δf has increased by 2. To get close to 75 kHz, we need to take the 780th harmonic. This cannot be done in one go but can be achieved by $3 \times 13 \times 20$. In order to generate harmonics, an amplifier can be driven to saturation and a tuned load used to select the wanted harmonic. Frequency shifting is required after the harmonics have been generated. In Figure 4.21, the carrier frequency after harmonic generation is 15.6 MHz. To get to 100 MHz, the signal will need to be shifted by 84.4 MHz. If this is done in one go, the sum and difference products will be 100 and 68.8 MHz. These might be too close together to adequately filter them using a high-frequency tuned circuit. (Recall that high-frequency tuned circuits have a low Q and a large bandwidth and so might not be selective enough.) The solution is to use two frequency shifters such as 20 MHz (frequencies of 35.6 and 15.6 MHz) and 64.4 MHz (frequencies of 100 and 28.8 MHz).

An alternative to the varicap oscillator is to use a crystal-controlled PLL. This has the advantage of being highly stable, unlike the LC oscillator just considered, and it is not restricted to narrowband FM. Figure 4.22 shows a block diagram of an FM modulator using a PLL. Let us take the loop operating with no modulation applied to the VCO and let the VCO run at a nominal frequency of 100 MHz. The output of the

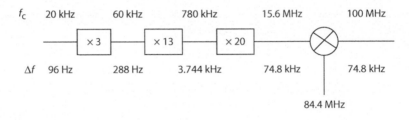

FIGURE 4.21 Frequency and harmonic generation for an FM transmitter.

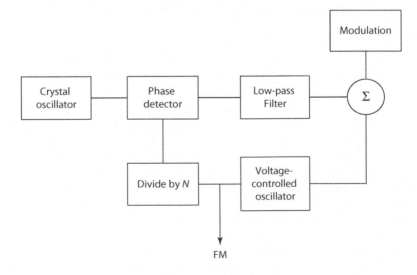

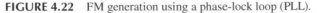

FIGURE 4.22 FM generation using a phase-lock loop (PLL).

VCO is divided by N (let's say $N = 100$). Thus, the frequency presented to the phase comparator is 1 MHz. Let the crystal oscillator be running at 1 MHz also. Any differences in phase, and frequency, will cause a dc voltage to be generated such that the phase error is in fact 90°. (If the phase error is 0°, the VCO voltage would be zero and phase shifts could cause a negative VCO voltage. By effectively biasing the VCO, this problem is avoided.) The loop is then in phase lock. The addition of the modulation to the VCO means that the frequency of the VCO varies according to the modulation. The advantage of this scheme is that it can run at the nominal carrier frequency directly and produce quite large carrier deviations. (A more detailed explanation of the PLL is given when we consider demodulation in Appendix XI.)

A further alternative is the Armstrong modulator. This design synthesises a narrowband FM signal. As shown in Figure 4.23, the modulation is first integrated with respect to time. This gives the inverse relationship with ω_m and the proportional relationship with V_m needed for the modulation index. The output of the integrator is mixed with the carrier to give

$$A = \frac{V_m}{\omega_m}\sin\omega_m t \times V_c \cos\omega_c t$$

$$= \frac{V_m V_c}{\omega_m}\left[\sin\left(\omega_c + \omega_m\right)t - \sin\left(\omega_c - \omega_m\right)t\right]$$

Therefore, the NBFM signal is given by

$$\text{NBFM} = V_c\sin\omega_c t + \frac{V_m V_c}{\omega_m}\left[\sin\left(\omega_c + \omega_m\right)t - \sin\left(\omega_c - \omega_m\right)t\right] \tag{4.37}$$

As shown in Figure 4.24, this has the spectrum of NBFM. NBFM is extremely useful because it has the minimal bandwidth of AM and the 100% power efficiency of FM. Although the tuned circuit modulator is not used due to problems with frequency stability, it does show how the maximum frequency deviation comes about.

In the FM receiver, a limiting amplifier removes any amplitude variations in the FM signal to remove some aspects of noise. In addition, since it is the frequency that is being varied, there is no need to worry about the amplitude. Following the limiter, demodulation of FM is more usually done with a PLL operating at the receiver IF (normally 10.7 MHz). The block diagram of a typical PLL demodulator is shown in Figure 4.25.

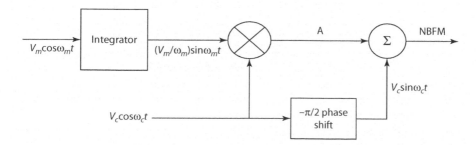

FIGURE 4.23 The Armstrong method of generating NBFM.

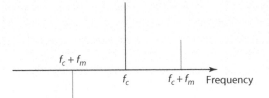

FIGURE 4.24   Spectrum of Armstrong NBFM modulator.

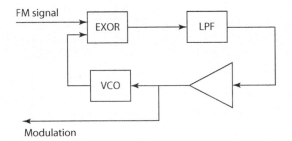

FIGURE 4.25   Block diagram of a phase-lock loop FM demodulator.

The first stage is the phase comparator, which can be as simple as an EXOR gate. (The limiter can give logic levels to interface with the PLL.) The input and the output signal of the VCO are compared. If the two signals are of different frequencies, the EXOR gate will generate a complex signal (Figure 4.26). This signal is smoothed in the LPF to produce an error voltage which is amplified and then applied to the VCO. The polarity of this voltage is such that the VCO adjusts itself to minimise the phase shift and hence error voltage. When the PLL is in lock, the input signal and the VCO are in the same phase and the error voltage is a fixed constant and the VCO voltage is maintained at a level needed to keep lock.

We have seen that in FM the frequency varies. When this is applied to a locked PLL, the VCO has to track the frequency changes. In order to do this, the voltage to the VCO has to alter in sympathy with the frequency transitions – the original modulation. Thus, the voltage to the VCO is the modulating signal. Care must be taken not to exceed the frequency range of the PLL. If the upper or lower frequencies of the VCO are exceeded, the loop will drop out of lock because it cannot follow the frequencies. The maximum minus minimum frequencies over which lock is maintained is called the *lock-in range*. Also of importance is the capture range which is the range of frequencies that the PLL will lock onto a signal. Appendix XI covers the mathematics relating to using a PLL as an FM demodulator.

The stereo broadcast signal for FM is more complicated than for AM. In AM, we only have one baseband signal as stereo is not used because of the high bandwidth requirement. In the FM stereo signal, there is a need for reception of the mono signal in case the stereo signal is not decoded due to noise or interference. Consequently, the FM baseband signal consists of three main components

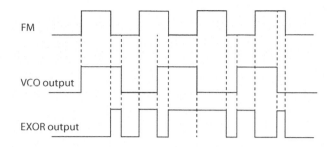

FIGURE 4.26 Timing diagram for a PLL.

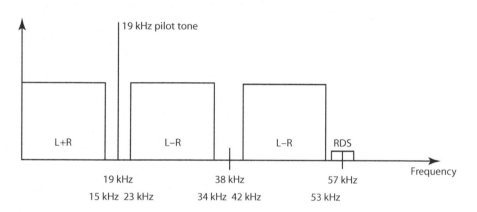

FIGURE 4.27 The stereo multiplex.

(Figure 4.27). The combining of the left (L) and right (R) channels prior to transmission is done in a multiplexer (Figure 4.28a) and the decoding back into L and R is done in a demultiplexer (Figure 4.28b). The audio is limited to the range 30 Hz to 15 kHz and the two L–R channels are a DSB-SC pair based at 38 kHz. The Radio Data Service (RDS in Europe) or Radio Broadcast Data Service (RBDS in the United States) is based around 57 kHz. Note that these frequencies are multiples of 19 kHz and they are phase locked to the pilot tone. The RDS/RBDS system transmits data at a rate of 1187.5 bit/s. One of the very useful services provided by RDS/RBDS is the automatic retuning of the receiver. Adjacent FM transmitters use different frequencies and so the user has to retune if he or she moves from one area to another. This can be dangerous if driving a car. A receiver equipped with RDS/RBDS can retune automatically and also retune for local traffic programmes as well. In the United States, additional services operate at frequencies above the RBDS. These serve a variety of uses such as talking newspapers for the blind, and private data transmissions. One thing that must be noted is that the noise on high frequencies within the FM multiplex is very large (as we will see next). Thus, special measures are needed to improve the S/N. Most countries use a 100 or 200 kHz channel spacing with a maximum frequency deviation of 75 kHz.

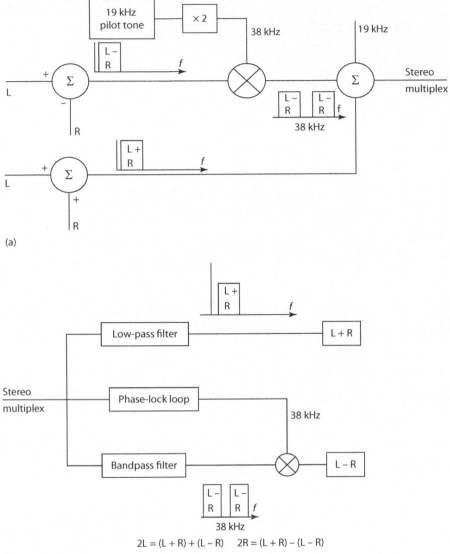

FIGURE 4.28 (a) Block diagram of an FM stereo multiplexer; (b) block diagram of an FM stereo demultiplexer.

We have seen that it is the in-phase component of noise that is significant in AM. Here, we consider FM and find that it is the quadrature component that is significant.

As shown in Equation 4.31, reproduced here in its sine wave form as Equation 4.36, the FM signal can be represented as a sine wave with a varying phase angle which depends on the modulating signal. Thus, it is the angle that is important and not the amplitude of the signal.

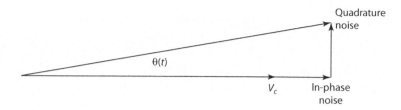

FIGURE 4.29 FM signal with in-phase and quadrature noise but no modulation.

$$v_{\text{FM}}\left(t\right) = V_c \sin\left(\omega_{\text{IF}} + \frac{k_f V_m}{f_m}\sin\omega_m t\right)t \tag{4.38}$$

Consider now the in-phase and quadrature noise components. As shown in Figure 4.29, the in-phase component adds to the carrier amplitude. A limiter is used before the detector to remove changes in amplitude and so it is the quadrature noise that affects the phase and hence frequency of the FM signal. The noise creates a random angle $\theta(t)$ which acts in a similar way to the modulation and so is demodulated in the same way. This angle is found as follows:

$$\tan\theta(t) = \frac{n(t)}{V_c + n(t)} \tag{4.39}$$

If the carrier is much greater than the noise (generally true but not always so as we will see):

$$\tan\theta(t) \approx \frac{n(t)}{V_c}$$

As the noise is much less than the carrier, the angle and its tangent are approximately equal. Thus,

$$\theta(t) \approx \frac{n(t)}{V_c} \tag{4.40}$$

where

$$n(t) = \sqrt{2\eta df}\,\cos\omega t$$

and ω is the frequency in the passband of the audio stage (previously ω_{off}). This random phase shift gives rise to a random frequency offset which is demodulated and appears as noise in the audio band. To convert the phase shift to a frequency, we need to differentiate Equation 4.40 with respect to time to give

$$\frac{d\theta}{dt} = \frac{\sqrt{2\eta df}\,\cos\omega t}{V_c}$$

$$= -\frac{\sqrt{2\eta df}\,\omega\sin\omega t}{V_c} \tag{4.41}$$

This frequency appears at the output of the detector in a similar way to the modulation. So, the noise power in a strip (similar to the AM case) is

$$\text{Noise power} = \left(\frac{\sqrt{2\eta df}\,\omega}{\sqrt{2}V_c}\right)^2 \Bigg/ R$$

$$= \frac{\eta df\,\omega^2}{V_c^2 R} \tag{4.42}$$

$$= \frac{\eta(2\pi)^2 f^2 df}{V_c^2 R}$$

Before we move on from Equation 4.42, it is worth noting that the noise power is proportional to the square of frequency. This causes a problem for high-frequency signals, as we will see next.

Equation 4.42 describes the noise power in a strip of frequency f at the output of the detector, i.e. at the input to the audio amplifier. To find the total noise power within the audio bandwidth, f_o, we need to integrate Equation 4.42 with respect to frequency with the limits $-f_o$ and $+f_o$. So,

$$\text{Audio noise power} = \int_{-f_o}^{+f_o} \frac{\eta(2\pi)^2 f^2 df}{V_c^2 R}$$

$$= \frac{\eta(2\pi)^2}{V_c^2 R} \int_{-f_o}^{+f_o} f^2 df \tag{4.43}$$

$$= \frac{\eta(2\pi)^2}{V_c^2 R}\left(\frac{2f_o^3}{3}\right)$$

This is the noise power in the audio bandwidth. To find the S/N, we need the audio power after the detector. The instantaneous frequency of the FM signal is

$$f = f_{\text{IF}} + k_f.V_m\cos\omega_m t$$

The detector will remove the IF to leave the audio as $k_f V_m \cos w_m t$. Converting to radian per second (rad/s) and calculating the power gives

$$\text{Signal} = \left(\frac{2\pi k_f V_m}{\sqrt{2}} \right)^2 \Bigg/ R$$

The audio S/N is hence

$$\frac{\text{Signal}}{\text{Noise}} = \left(\frac{2\pi k_f V_m}{\sqrt{2}} \right)^2 \Bigg/ R \Bigg/ \frac{V_c^2 R}{(2\pi)^2} \Bigg/ \frac{3}{2 f_o^3}$$

$$= \frac{3}{2} \frac{V_c^2}{2} \frac{1}{\eta f_o} \left(\frac{\Delta f}{f_o} \right)^2 \qquad\qquad (4.44)$$

$$= \frac{3}{2} \frac{C}{\eta f_o} \left(\frac{\Delta f}{f_o} \right)^2$$

where:
C is the carrier power
η is the noise power spectral density
f_o is the bandwidth of the following audio stage
Δf is the maximum frequency deviation of the FM signal

Taking an example, the signal voltage across the antenna terminals is 1 mV into 50 Ω giving a carrier power of 20 nW. We will take a noise power with spectral density of 8.5×10^{-20} W/Hz, a Δf of 75 kHz and an f_o of 15 kHz. Thus, Equation 4.44 becomes

$$\frac{3}{2} \frac{C}{\eta f_o} \left(\frac{\Delta f}{f_o} \right)^2 = \frac{3}{2} \frac{20 \times 10^{-9}}{8.5 \times 10^{-20} \times 75 \times 10^3} \left(\frac{75 \times 10^3}{15 \times 10^3} \right)^2$$

$$= 117 \times 10^6$$

$$= 80.7 \text{ dB}$$

This S/N is well above the level required for high-quality audio and shows the superiority of FM over AM. There is a problem with FM though. Recall from Equation 4.42 that the noise power increases as f^2 and so high-frequency audio signals suffer from more noise. Unfortunately, the signal power in high-frequency audio is low (it is mostly made up of harmonics) and so this f^2 dependency has a major effect. To counter this, a circuit called a *pre-emphasis circuit* is used to boost the high frequencies.

Figure 4.30 shows a simple pre-emphasis circuit with its associated frequency response. At low frequencies, the circuit acts as a potential divider and so the circuit attenuates low frequencies. When the reactance of the capacitor equals that of R_1, the capacitor begins to shunt R_1 and the signal attenuation gets less. There comes a point where the reactance of the capacitor equals that of R_2 and then the signal appears to pass straight through to the output. The frequency response is also shown

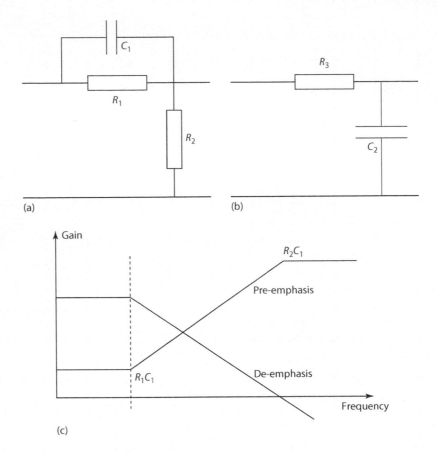

FIGURE 4.30 (a) Pre-emphasis, (b) de-emphasis networks and (c) associated frequency response.

in Figure 4.30. In the United Kingdom and Europe, the R_1C_1 time constant is 50 μs (3.18 kHz), while in the United States it is 75 μs (2.12 kHz). The time constant for R_2C_1 is chosen to be small enough not to affect the audio.

The pre-emphasis network affects the transmit signal whereas the de-emphasis affects the demodulated signal before the audio amplifier. So, the de-emphasis network will affect the noise as well. As observed previously, the noise power spectral density at the input to the audio amplifier has an f^2 characteristic as follows:

$$\text{Noise power spectral density} = \frac{\eta(2\pi)^2 f^2}{V_c^2 R} \text{ W/Hz} \tag{4.45}$$

The de-emphasis network is simply an LPF with transfer function:

$$\frac{1}{1 + j\omega RC} = \frac{1}{1 + j\frac{\omega}{\omega_c}} \tag{4.46}$$

where ω_c is the -3 dB frequency of the network. This transfer function is for voltages but we need it for power. The power transfer function is easily found by squaring Equation 4.46 to give

$$\left(\frac{1}{1 + j\,\omega/\omega_c} \right)^2$$

Multiplication by the noise power spectral density gives

$$\left(\frac{1}{1 + j\,\omega/\omega_c} \right)^2 a\omega^2 \tag{4.47}$$

where a is $\eta / V_c^2 R$. In order to simplify Equation 4.47, we define a frequency variable, f, as frequency normalised to the filter -3 dB frequency. Thus, ω becomes $f\omega_c$ and Equation 4.47 becomes

$$\left(\frac{1}{1 + jf} \right)^2 af^2\omega_c^2$$
$$= \frac{1}{1 + 2jf - f^2} af^2\omega_c^2 \tag{4.48}$$

The magnitude of Equation 4.48 is given by

$$\frac{1}{\sqrt{\left(1 - f^2\right)^2 + 4f^2}} af^2\omega_c^2 = \frac{1}{\sqrt{1 + 2f^2 + f^4}} af^2\omega_c^2$$
$$= \frac{af^2\omega_c^2}{1 + f^2} \tag{4.49}$$

Figure 4.31 shows the f^2 noise present at the input to the de-emphasis network and the noise at the output of the network. As can be clearly seen, the filtering effect has a major impact on the noise. To find the improvement in S/N, we need to integrate Equation 4.49 with respect to the normalised frequency and compare it to the unfiltered noise. The bandwidth we need to integrate over is ω_m or $f_m = \omega_m/\omega_c$. The time constant for Europe gives $f_c = 3.18$ kHz and, taking $f_m = 15$ kHz, we get $f_m = 4.72$. Therefore, the noise after the de-emphasis network is

$$\frac{\eta}{V_c^2 R} \omega_c^2 \int_0^{f_m} \frac{f^2}{1 + f^2}\, df = \frac{\eta}{V_c^2 R} \omega_c^2 \left(f_m - \tan^{-1} f_m \right) \tag{4.50}$$

The noise spectral density without the de-emphasis network is given by Equation 4.45 and so the noise at the input to the audio stage is

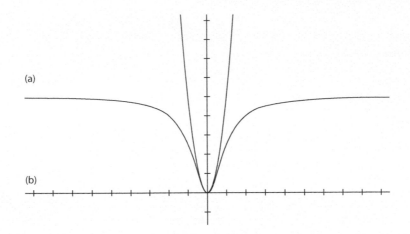

FIGURE 4.31   (a) f^2 noise at the input to a de-emphasis network and (b) noise after de-emphasis.

$$\frac{\eta}{V_c^2 R}\int_0^{\omega_m}\omega^2 d\omega = \frac{\eta}{V_c^2 R}\omega_c^2\int_0^{f_m}f^2 df$$

$$= \frac{\eta}{V_c^2 R}\omega_c^2\frac{f_m^{\,3}}{3}$$

(4.51)

The improvement is simply the ratio of Equation 4.51 to Equation 4.50 and, with the parameters used, the improvement is 10.2 dB.

There is a threshold in FM reception below which the approximations we have used to find the S/N do not hold. It happens when the noise and the carrier power are of similar order and it occurs when the C/N is approximately 7 dB or 5 as a ratio. This is very low and only occurs at the edge of the transmission region. If we are listening to a national radio station, there will be an alternative frequency that we could use (Figure 4.32). As shown in Figure 4.32, there are two regions that use the same frequency. This is known as *frequency reuse* and it is allowed because the regions are physically separated and VHF signals (where FM is transmitted) are attenuated as they pass through buildings, etc.

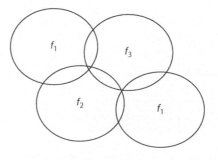

FIGURE 4.32   Frequency reuse with high frequencies.

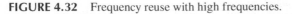

TABLE 4.1

Comparison of Broadcast AM and FM

	AM	FM
Carrier power (W)	20×10^{-9}	20×10^{-9}
Noise spectral density (W/Hz)	8.5×10^{-20}	8.5×10^{-20}
Modulation depth	100%	–
Maximum frequency deviation (Hz)	–	75×10^3
IF bandwidth (kHz)	50×10^3	200×10^3
Audio bandwidth (kHz)	15 kHz	15 kHz
S/N (dB)	4.7×10^6 (66.7 dB)	5.9×10^8 (87.8 dB)

Before we leave FM, a numeric example will show the advantage of FM over AM (Table 4.1). Both transmission formats have the same carrier power which comes from a terminal voltage of 1 mV into 50 Ω. They also have the same noise spectral density but different IF bandwidths because FM has a higher transmission band-width. The S/N using the previous work is 66.7 dB for AM and 87.8 dB for FM. This clearly demonstrates an FM improvement which can translate to longer transmission distances. Care must be taken when concluding this because FM is transmitted on VHF which has a limited transmission range. We can say, however, that FM has a better audio S/N than AM.

4.5 PHASE MODULATION

In FM it is the frequency of the carrier that varies. In PM it is the phase of the carrier that is varied according to the modulation.

Consider a carrier given by

$$v_c(t) = V_c \cos(\omega_c t + \theta(t)) \tag{4.52}$$

We have already discussed what happens if the amplitude and frequency of the carrier are varied according to the modulating signal. Here, we are concerned with phase variations, i.e. $\theta(t)$ in Equation 4.52 varies. Taking a phase constant of propor-tionality $k\theta$ rad/V, we get

$$\theta(t) = k_\theta V_m \cos \omega_m t \tag{4.53}$$

Substituting Equation 4.53 into Equation 4.52 yields

$$v_{PM}(t) = \cos(\omega_c t + k_\theta V_m \cos \omega_m t) \tag{4.54}$$

The instantaneous frequency, ω_i, is found by differentiating the phase term in Equation 4.54 with respect to t to give

$$\omega_i = \frac{d}{dt}\,\varphi(t) = \frac{d}{dt}\left(\omega_c t + k_\theta V_m \cos\omega_m t\right)$$

$$= \omega_c - k_\theta V_m \omega_m \sin\omega_m t$$

Thus,

$$v_{PM}(t) = V_c \cos\left(\omega_c - k_\theta V_m \omega_m \cos\omega_m t\right)t$$
$$= V_c \cos\left(\omega_c - m_\theta \cos\omega_m t\right)t \tag{4.55}$$

where

$$m_\theta = k_\theta V_m \omega_m \tag{4.56}$$

Compare this to the FM case (Equation 4.30):

$$m_f = \frac{k_f V_m}{f_m} \tag{4.57}$$

It is evident from Equation 4.56 that the PM index, $m\theta$, is directly dependent on the frequency of the modulation: double f_m and $m\theta$ doubles also. For FM, m_f is inversely dependent on the frequency of the modulation: double f_m and m_f halves. The bandwidth of both signals is highly dependent on the modulation indexes and the proportionality with f_m in PM means that the bandwidth will be larger than FM. This is why PM is not used for analogue modulation although it is used for digital modulation.

4.6   PROBLEMS

1. A 100 V, 10 MHz carrier is amplitude modulated by a 10 V, 1 kHz audio signal. Determine:
   a.   The maximum and minimum amplitudes of the modulated envelope.
   b.   The frequency of every component in the signal.
   c.   The amplitude of each component.
   d.   The modulation depth.
   e.   The ratio of the carrier power to the total power.
      [110 V, 90 V; 10.001 MHz, 10 MHz, 9.999 MHz; 5 V, 100 V, 5 V; 10%; 99.5%]
2. Show, by considering frequency components only, that a diode can be used as a modulator for AM. (Hint: you will need the dv³ term.)
3. In a coherent demodulator, the frequency is correct but there is a phase shift of 90° relative to the original carrier. Confirm that the baseband disappears when detecting AM and DSB-SC but that it is present in an SSB.

4. Confirm that the total power in an FM signal is equal to the power in the unmodulated carrier. (Square and sum the Bessel function coefficients for an m_f of 4 as an example.)

5. A narrowband FM signal is produced with a carrier of 1 MHz and $\Delta f = 5$ kHz. Determine suitable harmonic generation and mixing units to achieve wideband FM at a carrier of 98 MHz and a Δf of 75 kHz.

   [3rd and 5th harmonic, and shift of 83 MHz in total]

5 Transmission and Propagation of Electromagnetic Waves

In this chapter, we will discuss transmission lines, which are used to carry signals, and the propagation of electromagnetic waves which encompasses antennae.

Probably the most familiar example of a transmission line is the length of coaxial cable linking the TV antenna on the roof to the TV in the house. What is seldom realised is that a piece of wire is also a transmission line as is a printed circuit board (pcb) track. What is important is the relationship between the wavelength and the length of the line, and that is dealt with in this chapter. We must keep a sense of perspective when considering transmission lines. We will not have to worry about transmission line effects in a 10 m length of mains cable. The wavelength of 50 Hz in free space is 6×10^6 m and at 60 Hz it is 5×10^6 m; so, 10 m is a minute amount of wavelength. Conversely, if we take a 30 GHz signal and a 10 cm line, the line has 10 wavelengths on it and we would certainly need to use transmission line theory here.

There are many types of transmission lines – pcb track over a ground plane; twin feed (parallel conductors); 75 Ω coaxial cable (coax); 50 Ω coax; unshielded twisted pair (UTP); to name a few. Note that the choice of 75 Ω or 50 Ω coax is historical. It might be thought that there is little difference between the two; however, early experimentation showed that the best power handling capability came from 30 Ω coax whereas the lowest attenuation came from 77 Ω coax. So, 50 Ω was used as a compromise and it is used wherever power needs to be carried (transmitters and also radio receivers). TV receiver systems use 75 Ω and we will see why in Section 5.4.

5.1 WAVES ON TRANSMISSION LINES

Consider a transmitter connected to an antenna via a length of coaxial cable. When the transmitter is turned on, the signal travels down the cable to the antenna. This takes time – nothing happens instantaneously. To study this propagation, we need to consider the variation of voltage and current along the line.

Let us initially consider a lossless length of coaxial cable. (Other transmission lines can be treated in exactly the same way.) The line will have a series inductance associated with the magnetic field due to the current in the wire. There will also be a parallel capacitance associated with the potential difference between the inner and outer conductors. The inductance and capacitance are per unit length values (Appendix XII).

Consider a small section of cable of length Δx (Figure 5.1). We can use Kirchhoff's laws to give

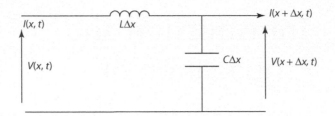

FIGURE 5.1 Section of a lossless transmission line.

$$v(x,t) - L\Delta x \frac{dI(x,t)}{dt} - \left(v(x,t) + \frac{dv(x,t)}{dx} \Delta x \right) = 0 \tag{5.1}$$

$$i(x,t) - C\Delta x \frac{dV(x+\Delta x,t)}{dt} - \left(i(x,t) + \frac{di(x,t)}{dx} \Delta x) \right) = 0 \tag{5.2}$$

By letting $\Delta x \to dx$, and ignoring second-order effects, Equations 5.1 and 5.2 become

$$-\frac{dv(x,t)}{dx} = L \frac{di(x,t)}{dt} \tag{5.3}$$

$$-\frac{di(x,t)}{dx} = C \frac{dv(x,t)}{dt} \tag{5.4}$$

An examination of Equations 5.3 and 5.4 reveals that they are linked – Equation 5.3 has *di/dt*, whereas Equation 5.4 has *di/dx*. So, if we differentiate Equation 5.3 with respect to *x* and Equation 5.4 with respect to *t*, we get

$$\frac{d^2v(x,t)}{dx^2} = -L \frac{d^2i(x,t)}{dxdt} \tag{5.5}$$

$$\frac{d^2i(x,t)}{dxdt} = -C \frac{d^2v(x,t)}{dt^2} \tag{5.6}$$

By substituting Equation 5.6 into Equation 5.5, we get

$$\frac{d^2v(x,t)}{dx^2} = LC \frac{d^2v(x,t)}{dt^2} \tag{5.7}$$

Similarly for the current:

$$\frac{d^2i(x,t)}{dx^2} = LC \frac{d^2i(x,t)}{dt^2} \tag{5.8}$$

Equations 5.7 and 5.8 are known as the wave equations and possible solutions are that of a travelling wave in the positive or negative direction, i.e.

$$v(x,t) = V_o \cos(\omega t \pm \beta x) \tag{5.9}$$

At first sight it is not immediately obvious whether the wave is going in a positive or a negative direction. In fact, the negative in Equation 5.9 indicates a positive travelling wave as we will see shortly. The parameter β is called the *phase coefficient* with units of radians per metre (rad/m) and to find it we need to substitute Equation 5.9 into the wave Equation 5.8. It is easier to do this if we use the phasor version of Equation 5.9. Using Euler's expression:

$$e^{j\varnothing} = \cos\varnothing + j\sin\varnothing$$

and so Equation 5.9 becomes

$$v(x,t) = V_o \cos(\omega t - \beta x)$$

$$= V_o e^{j(\omega t - \beta x)}$$

$$= V_o e^{j\omega t} e^{-j\beta x}$$

where we are considering the real part only. The advantage of this method is that differentiation is very simple.

$$\frac{dv(x,t)}{dt} = j\omega V_o e^{j\omega t} e^{-j\beta x} = j\omega v(x,t)$$

$$\frac{d^2 v(x,t)}{dt^2} = (j\omega)^2 V_o e^{j\omega t} e^{-j\beta x} = (j\omega)^2 v(x,t)$$

So, the voltage wave equation becomes

$$(-j\beta)^2 v(x,t) = LC(-j\omega)^2 v(x,t)$$

Hence,

$$\beta = \omega\sqrt{LC} \tag{5.10}$$

Figure 5.2 shows a plot of Equation 5.9 at time $t=0$ and at time $t=t_1$. A very important parameter can be defined first of all and that is the wavelength, λ. This is the distance over which the wave repeats itself. So,

$$\cos(\omega 0 - \beta\lambda) = 1$$

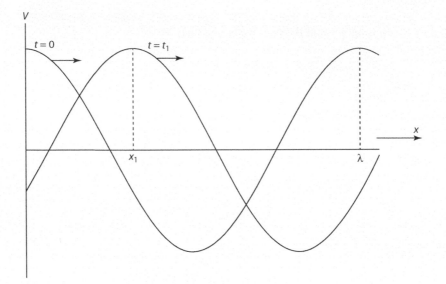

FIGURE 5.2   A wave travelling in the *x* direction.

Therefore, $\beta\lambda = 2\pi$, giving

$$\lambda = \frac{2\pi}{\beta} \tag{5.11}$$

The wave moves to the right – in time t_1 it moves a distance x_1 – thus it has a velocity in the positive *x* direction. To find this velocity, we note that the initial peak of the wave occurs at $x=0$, $t=0$ and that the same phase point occurs at distance x_1 at time t_1. Thus,

$$\cos(\omega t_1 - \beta x_1) = 1$$

Thus, $\omega t_1 - \beta x_1 = 0$, giving

$$\frac{x_1}{t_1} = v_p = \frac{\omega}{\beta} = \frac{\omega}{\omega\sqrt{LC}} = \frac{1}{\sqrt{LC}} \tag{5.12}$$

The parameter v_p is called the *phase velocity* and substitution of the expressions for inductance and capacitance gives the phase velocity as $1/\sqrt{\mu\varepsilon}$ where μ and ε are the permeability and permittivity, respectively, of the dielectric used in the transmission line. If the transmission line is air-cored, μ_o and ε_o are $4\pi\times10^{-7}$ H/m and 8.854×10^{-12} F/m and the phase velocity is the speed of light (3×10^8 m/s). With a denser, non-magnetic dielectric, the permeability is still that of free space, but the permittivity is increased by ε_r and so the phase velocity is lower. (Exactly the same effect occurs in glass.)

Every transmission line has an impedance – the characteristic impedance – but not the sort that we can measure using an ohm meter. This is the "surge" impedance and it is also the impedance presented to an electromagnetic wave. Like resistance, the characteristic impedance, Z_o, is V/I. By using Equation 5.3, we get

$$-\frac{dv(x,t)}{dx} = L\frac{di(x,t)}{dt}$$

$$-(-j\beta)v(x,t) = j\omega Li(x,t)$$

$$\frac{v(x,t)}{i(x,t)} = \frac{\omega L}{\beta}$$

$$Z_o = \frac{\omega L}{\omega\sqrt{LC}} = \sqrt{\frac{L}{C}} \tag{5.13}$$

If the transmission line is air-cored, we get $Z_o = 377\ \Omega$ which is the impedance of free space. Coaxial cable has a designed impedance of $50\ \Omega$ for radio and $75\ \Omega$ for TV.

If the line is imperfect, there will be some losses due to the resistance of the inner conductor (R) and any leakage from the inner to the outer conductor (a conductance of G). As shown in Figure 5.3, there are additional components – the series resistance, $R\Delta x$, and the parallel conductance, $G\Delta x$. By following a similar procedure to the lossless case, the solution to the voltage wave equation is

$$V(x,t) = V_o e^{j\omega t} e^{\pm \gamma x} \tag{5.14}$$

where γ is called the *propagation coefficient* which is defined as

$$\gamma = \alpha + j\beta \tag{5.15}$$

with α being the attenuation coefficient and β the phase coefficient as before. The propagation coefficient can be found from

$$\gamma^2 = (R + j\omega L)(G + j\omega C) \tag{5.16}$$

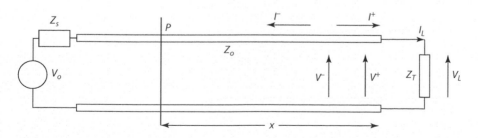

FIGURE 5.3 The lumped parameter model of a lossy transmission line.

With these definitions, for a wave travelling in the positive x direction, we can now write

$$V(x,t) = V_o e^{-\alpha x} e^{j\omega t} e^{-\beta x} \tag{5.17}$$

Note that we now have an exponential decay, meaning that the signal is attenuated as it goes down the line in the positive x direction. We should also note that the reactive components (C and L) dominate the propagation coefficient at high frequencies and so the line becomes almost lossless (Problem 2).

5.2   REFLECTIONS AND TRANSMISSION

The characteristic impedance is very important in communications. We have already seen when we considered noise that components in systems must be matched to minimise noise and also to transfer maximum power. Matching is vitally important to minimise reflections on transmission lines and in free space. As an example, consider light propagating in air ($Z_o = 377\ \Omega$) and let this light meet a pane of glass. The Z_o of glass is different to that of air and so there is a mismatch in impedance which causes some of the light to be reflected – there is transmission (we can see through the glass) and there is reflection from the glass. Two parameters, the reflection coefficient, Γ, and the transmission coefficient, τ, can be defined.

Consider a transmission line of characteristic impedance Z_o terminated in a load of Z_T, as shown in Figure 5.4. The incident voltage, V^+, is reflected off the load to give a reflected voltage, V^-. The reflection coefficient, Γ, is given by

$$\Gamma = \frac{V^-}{V^+}$$

Now, $V_L = V^+ + V^-$, and so

$$V^- = V_L - V^+$$

$$= I_L Z_T - V^+$$

$$= (I^+ - I^-)Z_T - V^+$$

$$= \left(\frac{V^+}{Z_o} - \frac{V^-}{Z_o}\right)Z_T - V^+$$

On rearranging,

$$V^-\left(1 + \frac{Z_T}{Z_o}\right) = V^+\left(\frac{Z_T}{Z_o} - 1\right)$$

Therefore, the reflection coefficient, Γ, is

$$\Gamma = \frac{V^-}{V^+} = \frac{Z_T - Z_o}{Z_T + Z_o} \tag{5.18}$$

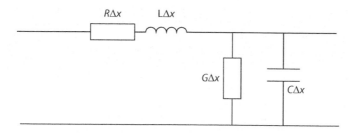

FIGURE 5.4 Incident and reflected signals on a mismatched transmission line.

Similarly, the transmission coefficient, τ, is

$$\tau = \frac{2Z_T}{Z_T + Z_o} \tag{5.19}$$

There are three loads of interest. If $Z_T=0$, i.e. a short circuit, $\Gamma=-1$ and all the incident signal is inverted and reflected. So, if the input is 10 V, the reflected voltage will be -10 V and the net voltage at the load will be the sum which is zero (as it should be for a short circuit). If the load is an open circuit, $\Gamma=+1$ and all the incident signal is reflected. So, if the input is 10 V, the reflected voltage will be $+10$ V and the net voltage at the load will be 20 V. Ideally, we want zero reflections and this is achieved when $Z_T=Z_o$, i.e. the load is matched to the line. Remember that this is also the condition for maximum power transfer.

Constructive and destructive interference (Figure 5.5) leads to a standing wave pattern. To find this pattern, we consider the combination of the incident and reflected waves. Let the line be lossless and the load be an open circuit. Figure 5.4 shows the situation. Let us take a point P at distance x from the load and let the voltage at this point be V_p. The wave incident at P, V_i, is simply a phase-delayed version of the source voltage, V_s, assuming negligible losses. So,

$$V_i = V_s \exp\left(-j\beta(l-x)\right) \tag{5.20}$$

where l is the length of the line. The reflected voltage at point P, taking an open circuit as the load, is

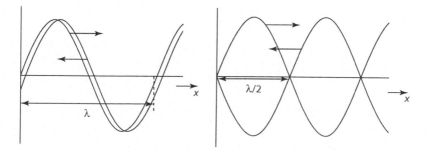

FIGURE 5.5 Constructive and destructive interference.

$$V_r = V_s \exp\left(-j\beta(l+x)\right)$$ (5.21)

The voltage at point P is the sum of these two voltages. So,

$$
\begin{aligned}
V_p &= V_i + V_r \\
&= V_s \exp\left(-j\beta(l-x)\right) + V_s \exp\left(-j\beta(l+x)\right) \\
&= V_s \exp\left(-j\beta l\right)\left[\exp\left(+j\beta x\right) + \exp\left(-j\beta x\right)\right] \\
&= V_s \exp\left(-j\beta l\right)\left[\cos\left(\beta x\right) + j\sin\left(\beta x\right) + \cos\left(\beta x\right) - j\sin\left(\beta x\right)\right] \\
&= 2V_s \exp\left(-j\beta l\right)\cos\left(\beta x\right)
\end{aligned}
$$ (5.22)

The term $V_s\exp(-j\beta l)$ is the source voltage, V_s, with a phase shift due to the length of the line, l. This is a constant and represents the voltage at the end of the line, V_l. The cos (βx) term generates the standing wave pattern. If $x=0$, i.e. we are at the load which is an open circuit, then the load voltage will be $2V_l$, i.e. twice the source voltage. If we work backwards, by $\lambda/4$ we get to zero volts (recall that $\beta=2\pi/\lambda$) and a pattern repeats itself as a peak at multiples $\lambda/2$ interspersed with nulls (Figure 5.6).

An important parameter when dealing with transmitting stations is the voltage standing wave ratio (VSWR). A standing wave results when the forward travelling wave meets the reflected wave, as shown in Figure 5.6. The VSWR is the ratio of the magnitude of the maximum to the minimum standing wave voltage along the line. The maximum voltage along the line is

$$
\begin{aligned}
\left|V_{\max}\right| &= \left|V^+\right| + \left|V^-\right| \\
&= \left|V^+\right| + \left|\Gamma\right|\left|V^+\right| \\
&= \left|V^+\right|\left(1+\left|\Gamma\right|\right)
\end{aligned}
$$ (5.23)

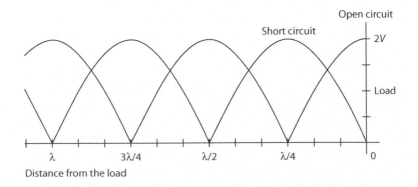

FIGURE 5.6   Standing wave plots for open-circuit and short-circuit loads.

The minimum voltage is

$$|V_{\min}| = |V^+| - |V^-|$$

$$= |V^+| - |\Gamma||V^+| \tag{5.24}$$

$$= |V^+|(1 - |\Gamma|)$$

Hence, the VSWR is

$$VSWR = \frac{|V_{\max}|}{|V_{\min}|} = \frac{1 + |\Gamma|}{1 - |\Gamma|} \tag{5.25}$$

The reflection coefficient varies from 0 to 1 in magnitude. A value of zero indicates a matched condition in that there are no reflected signals and so the VSWR is one – there is no standing wave pattern. If the reflection coefficient is 1 in magnitude, the VSWR is infinite. This is not correct because we have neglected the effects of attenuation down the line. This means that the reflected wave is smaller in amplitude than the incident wave because it travels further, and is attenuated more, having travelled to the load and back. Thus, the null voltage will be greater than zero. As can be seen from Figure 5.6, the distance between the nodes (peaks) is $\lambda/2$ which is the same for the anti-nodes (nulls). In practice, the VSWR can be measured using an SWR meter. The return loss is a useful parameter, defined as

$$\text{Return loss} = 20 \log |\Gamma| \tag{5.26}$$

So, the return loss is another measure of how much voltage is reflected.

Short-circuited and open-circuited stubs (short lengths of transmission line) are very useful elements for use in transmission lines. Depending on termination, they appear to be inductive or capacitive. Consider the situation in Figure 5.7, which is a simplified version of Figure 5.4. We have a length of transmission line and then a load. The source voltage, V_s, is

$$V_s = V^+ + V^- \exp(-j\beta 2l)$$

$$= V^+ + \Gamma V^+ \exp(-j\beta 2l)$$

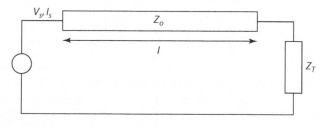

FIGURE 5.7 A simplified transmission line.

The source current, I_s, is

$$I_s = I^+ - I^- \exp(-j\beta 2l)$$

$$= \frac{V^+}{Z_o} - \frac{V^-}{Z_o} \exp(-j\beta 2l)$$

$$= \frac{V^+}{Z_o} - \Gamma \frac{V^+}{Z_o} \exp(-j\beta 2l)$$

Thus, the input impedance, Z_{in}, is

$$Z_{in} = \frac{V_s}{I_s} = Z_o \left(\frac{V^+ + \Gamma V^+ \exp(-j\beta 2l)}{V^+ - \Gamma V^+ \exp(-j\beta 2l)} \right)$$

$$= Z_o \left(\frac{1 + \Gamma \exp(-j\beta 2l)}{1 - \Gamma \exp(-j\beta 2l)} \right)$$

$$= Z_o \frac{\exp(-j\beta l)}{\exp(-j\beta l)} \left(\frac{\exp(+j\beta l) + \Gamma \exp(-j\beta l)}{\exp(+j\beta l) - \Gamma \exp(-j\beta l)} \right)$$

$$Z_{in} = Z_o \left(\frac{\exp(+j\beta l) + \Gamma \exp(-j\beta l)}{\exp(+j\beta l) - \Gamma \exp(-j\beta l)} \right)$$

This equation can be simplified by substituting for Γ from Equation 5.18 and expanding the exponentials into $\cos + j\sin$, to give

$$Z_{in} = Z_o \left(\frac{Z_T + jZ_o \tan \beta l}{Z_o + jZ_T \tan \beta l} \right) \tag{5.27}$$

Lengths of transmission line that are $\lambda/4$ long are of interest. If $l=\lambda/4$, the tan function in Equation 5.27 tends to infinity. So, the imaginary terms dominate and

$$Z_{in} = Z_o \left(\frac{jZ_o \tan \pi/2}{jZ_T \tan \pi/2} \right) \tag{5.28}$$

$$= \frac{Z_o^2}{Z_T}$$

Thus, the load impedance is transformed by the line that is $\lambda/4$ long. This is known as a *quarter wavelength transformer*. It is also interesting to see what happens when the load is an open circuit or a short circuit. If Z_T is infinite, Equation 5.27 becomes

$$Z_{open} = Z_o \left(\frac{Z_T}{jZ_T \tan \beta l} \right)$$

$$= Z_o \frac{1}{j \tan \beta l}$$

(5.29)

Similarly, the input impedance for a short-circuit termination is

$$Z_{short} = Z_o j \tan (\beta l)$$ (5.30)

So, both terminations result in reactive components. The open-circuit stub is capacitive and the short-circuit stub is inductive. Note that attenuation has been neglected because we are dealing with short lengths of transmission line. The results given in Equations 5.29 and 5.30 are highly dependent on wavelength. The tan function goes to infinity at $\pi/2$ and this occurs when the length of the stub is $\lambda/4$. The fact that a quarter wavelength stub can be capacitive or inductive is put to good use at microwave frequencies where a track on a pcb can be used for tuning. It should also be noted that the characteristic impedance of a line can be found by multiplying Equations 5.29 and 5.30, which gives

$$Z_o = \sqrt{Z_{open} Z_{short}}$$ (5.31)

It is interesting to examine what happens to pulses as they propagate down a transmission line. We can do this by drawing a lattice diagram (Figure 5.8). Consider a transmission line as in Figure 5.7, which has a characteristic impedance of 100 Ω terminating in an impedance of 20 Ω, with a source impedance of 5 Ω. We will take a pulsed source that generates a 1 ns wide pulse at a voltage, V_s, of 10 V and have a propagation delay of 10 ns down the transmission line. (A 10 ns delay in a cable with a phase velocity of 1.5×10^8 m/s is a length of 1.5 m and a 1 ns wide pulse corresponds to a data rate of 1 Gbit/s.) The first calculation is to find the amplitude of the pulse travelling down the line. The generator has an internal impedance of 5 Ω and an emf of 10 V. This generator sees the impedance of the line and not the load because it is 10 ns away. The voltage of the pulse going down the line is

$$V_p = \frac{10}{5+100} 100 = 9.52 \text{ V}$$

We also need to define two reflection coefficients: at the load for pulses from the line to the load, Γ_L, and at the source for pulses travelling back down the line towards the source, Γ_s.

$$\Gamma_L = \frac{Z_T - Z_o}{Z_T + Z_o} = \frac{20-100}{20+100} = -0.67$$

$$\Gamma_s = \frac{Z_s - Z_o}{Z_s + Z_o} = \frac{5-100}{5+100} = -0.90$$

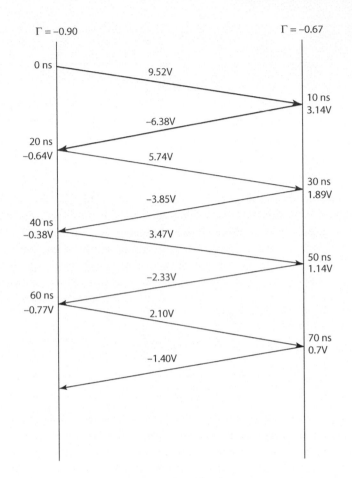

FIGURE 5.8 A lattice diagram showing pulse reflection.

As Figure 5.8 shows, the amplitude of the transient pulse reduces every time it gets reflected. It should also be noted that the initial pulse amplitude is reduced by the potential divider effect at the input. Thus, it is important to have sources with very low output impedance. The figures on the right-hand side (the load) show that the pulse amplitude never gets close to the original pulse. In fact, the highest amplitude is 3.14 V and this might not be enough to be useful to the load at the end of the line. If we let the pulse have a width much greater than the delay along the line (100 ns width say), we will have a steady-state situation. In this case, the terminal voltage is 8 V (potential divider between the source and load). To find the steady-state voltage from the lattice diagram, we simply add all the voltages on the right-hand side to give 6.87 V. Consideration of more reflections will give a closer result (see Problem 3). Note that if the line is terminated in the characteristic impedance of the line, there will not be a reflection from the load and the terminal voltage will be 9.52 V. This drop in voltage is unavoidable and is one reason why logic families register logic levels over a range of voltages.

5.3 SMITH CHARTS

In the previous section, we have seen how reflections down a line can cause power to return to the source. This is obviously undesirable and we would like to match the load to the line to prevent it. It is possible to produce an analytical solution, but it is far easier to use graphical methods, in particular the Smith chart (Figure 5.9).

The Smith chart uses load impedance normalised to the transmission line impedance, i.e. resistance is R/Z_o and reactance is X/Z_o. As an example, the normalised load impedance $1.5+j2$ is plotted in Figure 5.9 (Point A). If we have $1.5-j2$, it is in the lower half of the plot (Point B). As an exercise, it is useful to plot several points to become familiar with the chart. So, let us consider loads (antennae) on a 50 Ω transmission line of $100+j0$; $50-j10$; $10+j20$; $50+j0$ all of which have to be normalised to the line. So, the normalised impedances are 2; $1-j0.2$; $0.2+j0.4$; 1. These are plotted on the Smith chart of Figure 5.9 as C through F, respectively. Note that

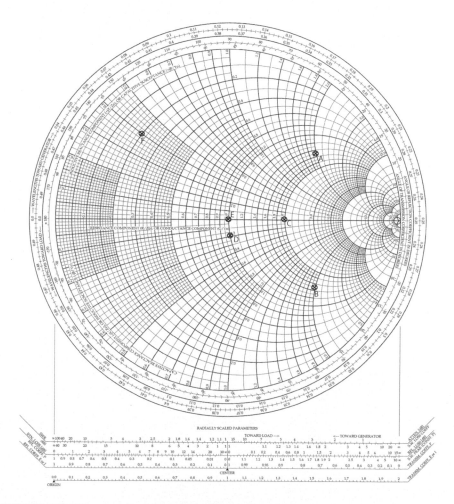

FIGURE 5.9 Six normalised impedances plotted on a Smith chart.

the last point, F, is the desirable matched condition – the load and the line have the same impedance.

Scales are drawn on the bottom of the chart. The bottom left-hand scale, labelled RFL COEFF E or I, is the reflection coefficient, Γ, which goes from unity at the left-hand side to zero in the centre. It is also worth noting that if Γ is one, the VSWR is infinite. This is because, as previously mentioned, the incident and reflected waves will have the same magnitude and so cancel each other out to give $V_{min}=0$ and an infinite VSWR. The VSWR can be readily found from the first left-hand scale. Impedances are plotted using the real – the horizontal axis that passes through the centre of the plot – and reactive components.

Let us consider another example. Let us take a section of a 50 Ω transmission line and a load of $30-j40$ Ω which represents an antenna. The normalised impedance is $0.6-j0.8$ and is plotted on the Smith chart in Figure 5.10. A circle is drawn with its centre being the centre of the chart (1.0) and radius that of the normalised load. This

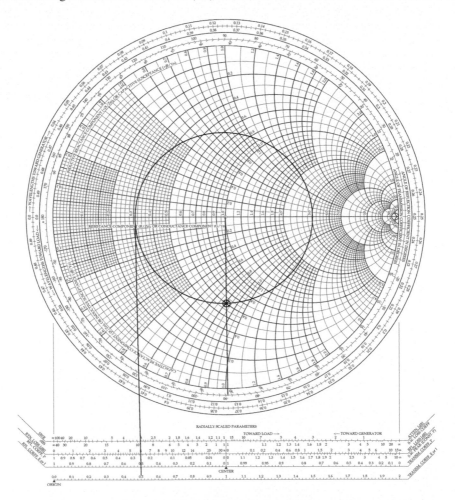

FIGURE 5.10 A plot of the normalised impedance $0.6-j0.8$.

circle is known as the VSWR circle. There are two ways to find the VSWR: drop a vertical line down to the first radial scale or read off the VSWR directly from the horizontal scale in the plot. Both result in a VSWR of 3.0. Another interesting feature is the reflection coefficient, Γ. The magnitude of this can be read off the scale as 0.5. (A check can be made using Equation 5.25 which relates the VSWR to the reflection coefficient.) The angle of Γ can be found by drawing a line from the centre of the chart to the load $0.6 - j0.8$ and continuing until the line meets the circular scale titled "Angle of reflection coefficient in degrees". For our load this gives an angle of 89°.

Thus far, we have used the Smith chart to find the VSWR and the reflection coefficient. We can use the chart to study what happens to the input impedance of the line as the length of the line varies. Consider a load of $100 + j150$ and a 75 Ω transmission line. As before, we calculate the normalised load to be $1.33 + j2$; plot this on a Smith chart (Figure 5.11) and draw the VSWR circle. The VSWR is 5 (either using the VSWR scale along the bottom or where the VSWR plot meets the horizontal line)

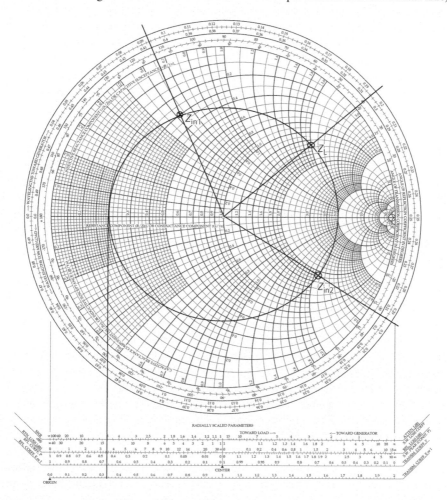

FIGURE 5.11   Smith chart for load of $100 + j150$.

and the reflection coefficient is 0.67/40°. We find the input impedance of the line at a point 0.4λ (say) by adding 0.4λ to the "wavelength" of the load, which is 0.194λ, to give 0.594λ. To find the point on the VSWR plot, we start at the far left-hand side (marked as 0.0) and use the "wavelengths towards generator" scale and rotate by 0.594λ. One complete rotation is 0.5 wavelengths and so we really only need to rotate by 0.094λ to give us $Z_{in1} = 0.3 + j0.64$ or $22.5 + j48$ Ω. If the line is 0.6λ long, we rotate by 0.794λ, which is $0.794 - 0.5 = 0.294\lambda$ from zero. This is marked as Z_{in2} on the chart and has the value $(1.8 - j2.2) \times 75$ Ω $= 135 - j165$ Ω. From this example, it should be evident that the length of the line has an effect on the load presented to the generator and this will affect the reflection coefficient and the VSWR.

Matching is a very important part of transmitter design. If the load (antenna) is not matched to the source, maximum power transfer does not take place and power is reflected back to the transmitter. It is possible to use shunt transmission lines to match the load and transmission line such that there are no reflections. Consider a load of normalised impedance $0.5 - j0.3$ connected to a 50 Ω transmission line which is connected to a 50 Ω source. The normalised load is plotted on the Smith chart (Figure 5.12) and a line is drawn from the load to the centre. What we are trying to do is move along the VSWR circle until we reach the unity circle (marked with the second line). The angle we have to move is $71° - 33.5° = 37.5°$. This can be read from the Smith chart. Now, one rotation around the chart (360°) corresponds to $\lambda/2$ and so

$$37.5° = 37.5 \times \frac{\lambda}{2 \times 360} = 0.052\lambda$$

We are now at the point marked with the arrow on the circle where the normalised impedance is 1. We want to get to the centre where the reactance is zero and the normalised impedance is 1. The impedance at 0.052λ is easily read off as $1 - j0.95$ and so we need to compensate for the $-j0.95$ reactance with a positive inductive reactance of $0.95 \times 50 = 47.5$ Ω.

We already know that the input impedance of a transmission line depends on its length and on whether it is terminated in a short circuit or an open circuit. Such lines are reactive and so we can use them to match to a load. Consider a load of $20 - j30$ connected to a 50 Ω transmission line with a stub connected, as in Figure 5.13. As this is a parallel combination, it is more usual to work with admittances. The normalised load is $0.4 - j0.6$ and this is plotted as Point A on the Smith chart in Figure 5.14. As normal, a circle is drawn through Point A. A line is then drawn through the centre, and Point B at the intersection of the line and the circle is the admittance. (A simple check can be carried out by taking the reciprocal of the normalised load and confirming from the Smith chart.) We require to be on the unity VSWR circle as before, so we draw a circle for one. As can be seen, there are two possible solutions which give us a VSWR of unity – Points C and D. Taking Point C first, the distance from the load is $0.174\lambda - 0.155\lambda = 0.019\lambda$. This is very close to the load and may not be practical. In spite of this, we will continue as it will be instructive when we consider Point D. As before, we require $Z_{in} = 1$ and so $y_{in} = 1$. Now, y_{in} is made up of the admittance of the transmission line to the right of the stub, y_x, and the admittance of the stub itself, y_l.

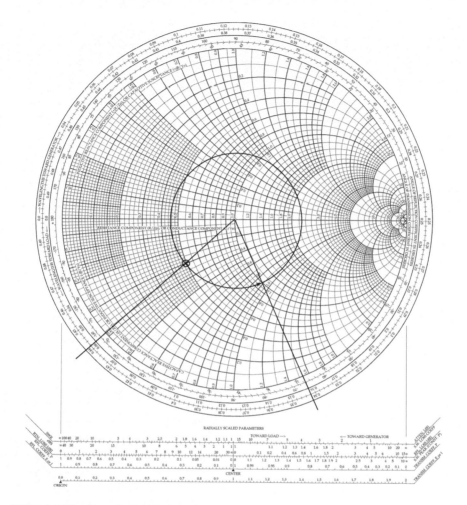

FIGURE 5.12 Use of a Smith chart for stub matching.

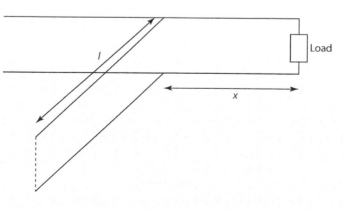

FIGURE 5.13 Stub matching.

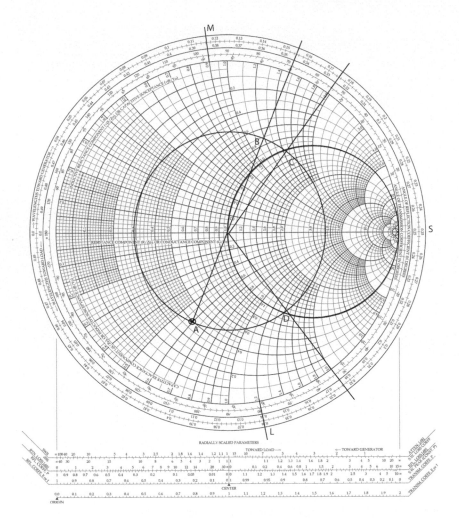

FIGURE 5.14   Smith chart for single stub matching.

So, $y_{in} = y_x + y_l$

Therefore, $1 = 1 + j1.2 + y_l$

And so, $y_l = -j1.2$

Hence, to match using the stub, the stub should be 0.019 wavelengths from the load and should have an admittance of $-j1.2$. In order to determine the length of the stub, we note that the normalised admittance of a short circuit is infinity which is marked as Point S on the far right-hand side of the Smith chart. We then move around the chart until we reach the line for 1.2 reactive – Point L. The stub length is the "distance" between S and L, which is $0.36\lambda - 0.25\lambda = 0.11\lambda$. So, for Point C the stub should be 0.11λ and placed at 0.019λ from the load.

For the alternative Point D, we follow a similar procedure. The distance from the load is $0.36\lambda - 0.155\lambda = 0.205\lambda$. As before, we have

$$y_{in} = y_x + y_l$$

Therefore, $1 = 1 - j0.9 + y_l$

And so, $y_l = +j0.9$

In order to determine the length of the stub, we move around the chart from Point S in an anticlockwise direction (towards the generator) until we reach the line for 0.9 reactive – Point M. The stub length is the "distance" between S and M, which is $0.384\lambda - 0.25\lambda = 0.134\lambda$. So, for Point D the stub should be 0.134λ long and placed at 0.205λ from the load.

5.4 ANTENNAE

Antennae are an essential part of a communications link, be it point to point or broadcast, transmitter or receiver. A receiving antenna can be a simple piece of wire, a telescopic antenna or a complex system of dipole and directors. The transmitting antenna is more complex because, as we have seen, matching the source and the load is very important. Rather than analyse various antennae from a theoretical point of view, we will limit our discussion to a qualitative analysis.

On the transmitter side, power travels up the feeder and into the antenna. There is usually a matching network which matches the antenna to the feeder. This is the purpose of the balun, dealt with shortly. The antenna appears to be a load which consumes power so there is current (magnetic field) and voltage (electric field). The electromagnetic field is radiated from the antenna in a particular direction.

A reference antenna can be defined as an isotropic radiator – a point source of electromagnetic energy. As a point source, radiation occurs in all directions rather like the electric flux that radiates from a positive point charge. If we place a reflecting dish behind this point source and place the source at the focal point of the dish, we have focused the energy in the direction of the dish. This effectively gives us gain over the isotropic radiator.

As mentioned in Chapter 1, the optimum size of an antenna is $\lambda/2$ and the most common antenna is the half-wave dipole shown in Figure 5.15. Also shown in Figure 5.15 is the current distribution along the dipole. Note that the dipole will resonate when driven with a signal whose 1/2 wavelength matches the length of the dipole. There is a problem when driving a dipole and that is the feeder needs to be balanced so that

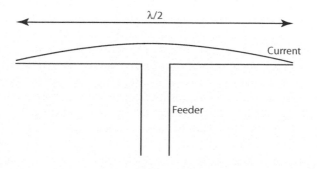

FIGURE 5.15   Current distribution in a half-wave dipole.

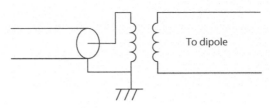

FIGURE 5.16   1:1 balun feeding a dipole.

both sides are fed equally and not referenced to ground. Coaxial cable cannot do this because one conductor (the sheath) is earthed, but twin feeder can. So, what is used in both transmitters and receivers is a balun (balance to unbalanced). A simple balun can be made from a 1:1 isolation transformer (Figure 5.16). The coaxial cable feeds the primary and the secondary is connected to the dipole. There are many different types of balun and we do not have space to discuss them here. What they all have in common is the transformation from unbalanced to balanced.

The impedance of a half-wave dipole is 73 Ω which is close to 75 Ω which is one of the standard impedances for coaxial cable. As the dipole is resonant there will be very little or no reactive component present. The polar plot of the dipole is shown in Figure 5.17. As can be seen, the dipole radiates in the forward and backward directions so there is some directivity. This is different to the isotropic radiator which radiates in all directions and so we can say that the dipole has some gain over the isotropic source of, specifically, 2.15 dB. The polar plot of Figure 5.17 is one of power variation with angle. As can be seen, there are two lobes present and maximum power is in two directions – 90° and 270°. As we move around the plot, the power reduces until it reaches zero at the centre. So, if we face the dipole, we will get maximum power in two directions whereas if we are alongside the dipole there is no power. The presence of the backward lobe indicates wasted power. It would be far better, and give more gain, if all of the power was concentrated in the forward direction. This is the purpose of a reflector which is placed half a wavelength behind the dipole. Directors can also be placed in front of the dipole and these operating together give high gain and high directivity. The resultant antenna is called a *Yagi–Uda antenna* or, more simply, a *Yagi antenna*. The design should be very familiar as it is used for very high frequency (VHF) and ultra high frequency (UHF) TV reception.

For mobile applications, we need an omnidirectional antenna. One such is the λ/4 monopole which acts as a λ/2 dipole (Figure 5.18) by virtue of the image conductor in the ground plane. The ground plane does not have to be fixed to the physical ground. Instead, a metal sheet, such as the roof of a car, can be used. The impedance of a λ/4 monopole is 36 Ω and its gain over an isotropic radiator is 5.14 dB. The radiation pattern is a circle so it receives and transmits signals from all horizontal directions. It is possible to change the impedance of the monopole to 50 Ω by using radial ground wires angled at 45° to the vertical or by using a transformer.

One last modification to the basic dipole is the folded dipole commonly used for TV reception. It is essentially a λ/2 dipole with the ends connected together

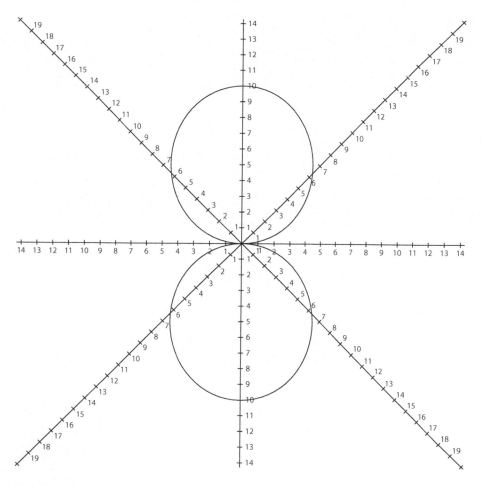

FIGURE 5.17 Polar plot of a dipole.

(Figure 5.19). Its impedance is nearly 300 Ω and so a balun must be used to get the 75 Ω necessary for coax.

A phased array consists of a number of transmitting antennae placed a certain distance apart and fed by the same signal but at different phases. Consider the two-element array shown in Figure 5.20. The two elements are spaced d metre apart and are fed with the same signal but with a phase shift of ϕ. At some point P in the far-field, i.e. far enough away that we have a plane wavefront, the two waves will either add constructively or destructively. The wave from Antenna 1 has an additional phase shift associated with it by virtue of the increased path length. With reference to Figure 5.20, this additional path length is $d \cos \theta$ and this gives rise to an additional phase shift of $\beta d \cos \theta$. If there is an additional electrical phase shift, α, the total phase difference between the two waves at point P is

$$\varphi = \beta d \cos \theta + \alpha$$

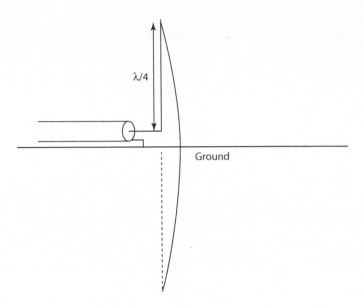

FIGURE 5.18 A quarter wave antenna.

where β is the phase coefficient ($\beta = 2\pi/\lambda$) or the phase shift per metre.
  The total electric field at point P is

$$E_P = E\left(1 + e^{j\varphi}\right)$$

$$= E e^{j\varphi/2}\left(e^{-j\varphi/2} + e^{j\varphi/2}\right)$$

  The magnitude of the $e^{j\varphi/2}$ term is one because it is $\sqrt{\cos^2 + \sin^2} = 1$, and so the magnitude of the electric field at P is

$$2E\cos\left(\varphi/2\right)$$

$$= 2E\cos\left(\frac{\beta d}{2}\cos\theta + \frac{\alpha}{2}\right) \tag{5.32}$$

FIGURE 5.19 A folded dipole.

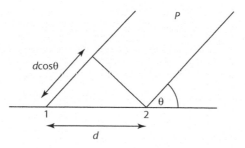

FIGURE 5.20   Phased array of two elements.

Let us initially examine when Equation 5.32 is a maximum and work with an antenna spacing, d, equal to λ. The maximum occurs when

$$\cos\left(\frac{\beta d}{2}\cos\theta + \frac{\alpha}{2}\right) = 1$$

and so,

$$\left(\pi\cos\theta + \frac{\alpha}{2}\right) = n\pi \tag{5.33}$$

where $n = 0, 2, 4$, etc. Taking $\alpha = 0$ gives

$$\cos\theta = n$$

The only solution is for $n = 0$ giving $\cos\theta = 0$ and so $\theta = 90°$. So, the maximum occurs at right angles to the array. If α is varied, the direction of the maximum field will change and so the beam can be steered to point in different directions. If the phase shift is varied dynamically, the beam will sweep out in front of the array and we would have a phased-array radar.

We have only studied isotropic radiators (point sources). To obtain the response due to an array of dipoles, we simply convolve the radiation pattern of a dipole with the radiation pattern of the isotropic radiators. It is possible to have more than two elements in the array. In this case, the E field at some point is

$$E_P = E\left(1 + e^{j\varphi} + e^{j2\varphi} + \cdots + e^{j(n+1)\varphi}\right)$$

This series can be summed to

$$E_P = E\left|\frac{1 - e^{jn\varphi}}{1 - e^{j\varphi}}\right|$$

And so,

$$E_P = E \left| \frac{\sin n\varphi/2}{\sin \varphi/2} \right| \tag{5.34}$$

The term in brackets is known as the *array factor* and it has a zero when $\varphi = 0$.

5.5   PROPAGATION

Depending on the wavelength, signals can cover a very large area or be highly local-ised. Low frequencies such as 200 kHz are used for broadcasting and can cover an area the size of the United Kingdom or a state. This propagation is due to diffraction of the wave as it passes over the earth. Slightly higher frequencies are attenuated and medium-wave transmission will often need two transmitters on different frequencies to cover the same area. There are problems with fading as night falls. This is due to the effect of the ionosphere. Next is short wave, which is an interesting band in that the signals are reflected off the ionosphere and are used for world service transmissions. Above short wave, we have frequency modulation (FM) transmissions as well as digi-tal radio on VHF. This is line of sight, and directional antennae are used to receive signals. On UHF, we have digital TV, mobile phones, Wi-Fi, etc. This is strictly line of sight and this is useful for frequency reuse. For transmission in the 10 GHz region, the ionosphere is transparent and signals can be sent to geostationary satellites for retransmission to the earth. The next chapter examines satellite communications.

Let us examine the effect of the ionosphere on short-wave signals. As the name suggests, the ionosphere consists of charged particles or ions. These ions are gener-ated by ultraviolet (UV) light coming from the sun and interacting with molecules in the atmosphere. The ionosphere is split into four layers – D, E, F_1 and F_2. The D layer extends from 60 to 90 km in a region of relatively dense atmosphere. UV radia-tion from the sun causes ionisation of the air. As this layer is close to the earth, the UV has to pass through a considerable amount of atmosphere. Thus, the intensity of the UV is quite low and this leads to a low level of ionisation. Recombination in this layer is also quite high because the atmosphere is relatively dense. The D layer is only present during the daytime.

The E layer extends from 90 to 150 km and is mainly due to the ionising effects of UV and x-rays. The atmosphere is more rarefied here and so there is not so much recombination. Thus, this layer reflects signals of around 10 MHz. This layer is pres-ent mainly during the day. There are two F layers – F_1 and F_2 – that are normally considered to be combined during the day. The F layer extends from 150 to 500 km and splits into two layers during the day – the period of high radiation. This layer is long lasting because the atmosphere is rarefied thereby leading to low recombina-tion. By using ionospheric propagation, signals can be sent over the horizon and can even be received on the other side of the world. It is primarily used by countries that have world service broadcasts – Voice of America; BBC World Service; Voice of Russia, etc. It is possible to mathematically analyse the reflection of signals from the ionosphere. We will do this by first finding the relative permittivity of a region of ionosphere and then we will use the result to find the condition for reflection.

Consider an electric field, E, from a transmitter located several hundred kilometres away from a point in the ionosphere. This electric field will cause any ions to move in the direction of the E field by virtue of Coulomb's law. We can also use Newton's second law here. Thus,

$$F = qE = ma \qquad (5.35)$$

If we consider the electric field to be a sinusoid, we can rewrite Equation 5.35 as

$$a = \frac{q}{m} E \cos \omega t \qquad (5.36)$$

Current flow is due to the movement of charge and so we can postulate that there is a current flow induced in the ionosphere by virtue of Equation 5.36. The current density, J Am^{-2}, is given by

$$J = Nqv \qquad (5.37)$$

where:
N is the ion density in cubic metres
v is the ion velocity in metres per second

The velocity of the carriers can be found by integrating Equation 5.36 to give

$$v = \frac{q}{m} \int E \cos \omega t \, dt$$

$$= \frac{q}{\omega m} E \sin \omega t$$

And so the current density (Equation 5.37) becomes

$$J = Nq \frac{q}{\omega m} E \sin \omega t \qquad (5.38)$$

The conduction current as given by Equation 5.38 is not the only current we have to consider. There is also the displacement current due to the "displacement" of carriers acting under the electric field. The displacement current density, dD/dt, is given by

$$\frac{dD}{dt} = \frac{d}{dt} \varepsilon_o E \cos \omega t$$
$$= -\omega \varepsilon_o E \sin \omega t \qquad (5.39)$$

where ε_o is the permittivity of free space. So, there are two components to the current in the ionosphere – the conduction current and the displacement current. (This is exactly the same as occurs in non-ideal capacitors.) We require the relative permittivity, ε_r. We can obtain this by equating the total current density to the displacement current density but using ε_r. Thus,

$$\frac{dD}{dt} = \frac{d}{dt}\varepsilon_o\varepsilon_r E \cos \omega t$$

$$= -\omega\varepsilon_o\varepsilon_r E \sin \omega t$$

This equals the sum of the conduction current density and the displacement current density. Thus,

$$-\omega\varepsilon_o\varepsilon_r E \sin \omega t = Nq\frac{q}{\omega m} E \sin \omega t - \omega\varepsilon_o E \sin \omega t$$

After cancelling $E \sin \omega t$ and rearranging, we get

$$\varepsilon_r = 1 - \frac{Nq^2}{\varepsilon_o \omega^2 m} \tag{5.40}$$

To calculate the angle at which the signal returns to earth, we assume an ideal reflecting surface. In reality, the ray is refracted and returns to earth but using an ideal reflector gives an indication of what happens. Figure 5.21 shows the situation to be analysed. In Figure 5.21, a ray hits the boundary between air and the ionosphere. If the angle is smaller than the critical angle, θ_c, the signal passes through the boundary, is refracted and is lost to space. If the angle is θ_c, the ray passes along the horizontal and if the angle is greater than θ_c the ray returns to earth. Snell's law of refraction shows

$$\frac{\sin \theta_i}{\sin \theta_r} = \frac{v_1}{v_2} = \frac{c/n_1}{c/n_2} \tag{5.41}$$

The refractive index of air is 1, so

$$\sin \theta_i = n_2 \sin \theta_r \tag{5.42}$$

The speed of light is given by $1/\sqrt{\mu_o\mu_r\varepsilon_o\varepsilon_r}$. We are dealing with a non-magnetic ionosphere and so μ_r will be one and the speed of light in the ionosphere will be $1/\sqrt{\mu_o\varepsilon_o\varepsilon_r}$. The speed of light in a vacuum is $1/\sqrt{\mu_o\varepsilon_o}$ and so the refractive index in the ionosphere, n_2, is $\sqrt{\varepsilon_r}$.

We already have an expression for ε_r in the form of Equation 5.40 and so Equation 5.42 becomes

$$\sin \theta_i = \sqrt{1 - \frac{Nq^2}{\varepsilon_o \omega^2 m}} \sin \theta_r$$

Thus,

$$\sin \theta_i = \sqrt{1 - \frac{81N}{f^2}} \sin \theta_r \tag{5.43}$$

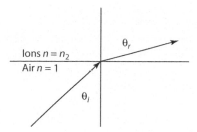

FIGURE 5.21   Refraction between air and the ionosphere.

The critical frequency, f_c, is defined as the frequency at which a vertically transmitted radio pulse gets reflected back to earth. This is important because, like a radar system, the height of a specific ionisation layer can be measured from the time taken for the pulse to return. Vertical incidence and reflection means $\sin \theta_i = \sin \theta_r = 0$, thus Equation 5.43 becomes

$$\frac{\sin \theta_i}{\sin \theta_r} = \sqrt{1 - \frac{81N}{f_c^2}} = 0$$

and the critical frequency is

$$f_c = 9\sqrt{N} \qquad (5.44)$$

Reflection off a layer occurs when θ_r is 90°. Thus, Equation 5.43 becomes

$$\sin \theta_i = \sqrt{1 - \frac{81N}{f^2}}$$

Therefore,

$$\sin^2 \theta_i = 1 - \frac{81N}{f^2}$$

As $1 - \sin^2$ is $\cos^2$, we get

$$\cos^2 \theta_i = \frac{81N}{f^2}$$

Thus,

$$\cos \theta_i = \frac{9\sqrt{N}}{f} \qquad (5.45)$$

This is the critical angle and so angles greater than θ_i will be reflected back to earth. The critical angle is highly dependent on the ion density and this is highly dependent on the sun. It is quite common for users to change their frequency of operation as night falls.

5.6 PROBLEMS

1. A telephone line has a length of 12 km and has the following primary coefficients per kilometre: $R = 120\ \Omega$, $L = 5$ mH, $C = 0.3\ \mu$F and $G = 30\ \mu$S. A sinusoidal signal with a frequency of 1194 Hz and amplitude 5 V is applied to the sending end of the line. Determine:
   a. The characteristic impedance of the line.
   b. The voltage and phase half way down the line if the line is terminated in its characteristic impedance.
   [$236/-36\ \Omega$; 0.746 V, $147°$]

2. Repeat Problem 1 with a frequency of 10 MHz. Comment on your result.
   [$129\ \Omega$; 5 V, $180°$. The line appears to be lossless]

3. With reference to Figure 5.7, continue calculating the steady-state terminal voltage up to 150 ns.
   [7.65 V]

4. Consider a load of $100 + j150\ \Omega$ connected to a 75 Ω transmission line. Using a Smith chart, determine the reflection coefficient, VSWR, the load admittance, the input impedance at a distance 0.4λ from the load and at the generator which is 0.6λ from the load. Confirm by calculation.
   [$0.66/40°$, 4.8, $3.2 \times 10^{-3} - j4.5 \times 10^{-3}$, $22.5 + j48.8$, $135 - j165$]

5. A line with a characteristic impedance of 100 Ω is terminated in a load of $20 + j50\ \Omega$. Determine the position and length of an open-circuit stub that will match the load.
   [0.076λ at 0.364λ from the load]

6. An array consists of four elements driven with a phase difference of $0°$ between them. Confirm that the maximum field occurs at right angles to the array. If $\alpha = -\beta d$, confirm that the maximum field occurs along the array.

7. A 10 MHz signal is to be reflected off an ionospheric layer that has an ion density of 3×10^{11} m^{-2}. Determine the critical angle and the skip distance if the signal is reflected off the layer at a height of 300 km.
   [$60°$, 1040 km]

6 Systems

In this chapter, we will discuss several real communications links in order to put some of the theory presented into real-world situations. Some of our discussions will be detailed whereas others will be more superficial. This is because whole books would be needed to adequately present some systems and we simply do not have the space.

6.1 SATELLITES

The ionosphere is transparent to microwave signals and this is put to good effect in satellite communications. Direct broadcast by satellite (DBS) is common place today with many TV and radio channels available to us. In common with terrestrial TV, satellite TV uses digital techniques to carry many channels – the DVB-S standard is used. This scheme uses a maximum 16-QAM and Reed–Solomon (RS) coding. In view of the high carrier frequency, the TV channel capacity is in excess of 200 channels.

In a satellite broadcast service, a ground station communicates with the orbiting satellite using microwave frequencies. The ionosphere that surrounds the earth is transparent at this frequency and communication is by direct line of sight. As we will see shortly, the signal power from the ground station can be quite large and that has an obvious advantage. Rather than describe an individual satellite system, we will discuss a generic link with realistic parameters.

Consider a geostationary satellite system. We will start at the ground station with an uplink frequency of 10 GHz and calculate the amount of power at the satellite 35,780 km away. Let the power from the modulator be 10 mW. Amplification of 40 dB gives a power of 100 W. The gain of a parabolic reflector is

$$\text{Gain} = \eta \left(\frac{\pi D}{\lambda} \right)^2 \tag{6.1}$$

where:
    η is the efficiency of the dish taken to be 80% (less than 100% due to the shadow of the arm holding the transmitter)
    D is the diameter of the dish (3 m)
    λ is the wavelength (3 cm)

$$\text{EIRP} = 10 \times 10^{-3} \times 1 \times 10^4 \times 7.9 \times 10^4 \, \text{W} = 7.9 \, \text{MW}$$

(The EIRP is the effective isotropic radiated power. Isotropic means a point source, which means that the power is radiated in all directions. So, the effective power is that required to generate the same power density [W/m²] at a distance from the source.) This power might seem rather excessive; however, it must be remembered that we have a dish that focuses energy into a fairly tight beam and this has the apparent effect of increasing the power. (A similar effect happens when a reflector is placed behind a light bulb. It appears brighter than it actually is.)

An isotropic radiator radiates in all directions by its very definition and so the power of 7.9 MW is from a point source giving a power flux density at a satellite 35,780 km away of

$$\text{Power} = \frac{7.9 \times 10^6}{4\pi \left(3.578 \times 10^7\right)^2} \text{ W/m}^2 = 500 \text{ pW/m}^2$$

We next need to multiply by the collecting area of the satellite dish to get the received power. An 80 cm dish has an effective area of 0.4 m² (assuming an efficiency of 0.8) and so the received power is

$$\text{Received power} = 500 \times 0.4 = 200 \text{ pW}$$

In Section 3.5, we examined what happens to the noise in a cascaded network, which we have here. Figure 6.1 shows the components of a satellite transponder. The antenna is connected to a low noise pre-amplifier which, in turn, is connected a high power post-amplifier. Filters are needed to limit the bandwidth to 30 MHz. The filter is not a simple resistor-capacitor (RC) filter because the frequency is too high. Instead, a physical resonator has to be used. After the filter, a mixer follows which converts the 10 GHz uplink frequency to the 12 GHz downlink frequency by mixing with 2 GHz. Filtering is used again to remove the unwanted 8 GHz component and a power amplifier generates the power needed for the downlink.

It should be noted that this transponder does not perform any demodulation and so it is important to have a powerful signal from the ground station. The effective noise temperature of the receiver antenna is very small because it is directional and does not "see" the whole of space. We will take a T_e of 50 K. The design of microwave components is quite advanced what with the use of gallium arsenide (GaAs) transistors and this means that gains are high and noise is low. The total gain is 110 dB giving an output power of 20 W.

As regards the noise, we can use Friss' formula (Section 3.4) and Table 6.1 to get

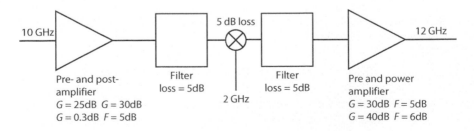

FIGURE 6.1   Block diagram of a satellite transponder.

TABLE 6.1

Gain and Noise of an Example Satellite Transponder

Component	Gain	Noise
Antenna	–	50 K
First pre-amplifier	25 dB = 316	1.5 dB = 1.41
Post-amplifier	30 dB = 1,000	5 dB = 3.16
Filter 1	−5 dB = 1/3.16	5 dB = 3.16
Mixer	−5 dB = 1/3.16	5 dB = 3.16
Filter 2	−5 dB = 1/3.16	5 dB = 3.16
Second pre-amplifier	30 dB = 1,000	5 dB = 3.16
Power amplifier	40 dB = 10,000	6 dB = 3.98

$$F_{tot} = 1 + (1.11 - 1) + \frac{(3.16 - 1)}{316} + \frac{(3.16 - 1)}{1000 \times 316} + \text{small}$$

$$= 1.17$$

The input noise from the antenna is

$$kTB = 2.1 \times 10^{-14} \text{ W}$$

where the bandwidth has been taken to be 30 MHz – large enough to accommodate many digital TV channels. So, the input signal to noise ratio (S/N) is

$$\frac{200 \times 10^{-12}}{2.1 \times 10^{-14}} = 9.7 \times 10^{3}$$

As $F = 1.11$, the output S/N is 8.7×10^3 or 39.4 dB. Remember that this is digital TV and so this S/N is quite respectable.

For the downlink, we need to calculate the EIRP at the downlink frequency (12 GHz). Taking an identical dish as for the uplink (80 cm dish with η of 0.80), the gain Equation 6.1 is

$$\text{Gain} = \eta \left(\frac{\pi D}{\lambda} \right)^2 = 8.1 \times 10^3$$

and so the EIRP for the downlink is

$$\text{EIRP} = 200 \times 10^{-12} \times 1 \times 10^{11} \times 8.1 \times 10^3 = 162 \text{ kW}$$

This power gives a power density at the customer premises of

$$\text{Power} = \frac{162 \times 10^3}{4\pi \left(3.578 \times 10^7 \right)^2} \text{ W/m}^2 = 10 \text{ pW/m}^2$$

And so the received power, taking a 1 m dish with 80% efficiency, is

$$10 \times 0.8 \times \pi (0.5)^2 = 6.3 \, \text{pW}$$

At the focal point of the dish is a unit called a *low-noise block* (LNB) converter which detects the signal, amplifies it and then mixes it to the range 950 MHz to 2 GHz. This frequency conversion is very important because carrying the satellite signal (12 GHz) to the set-top box would be difficult and expensive. (Conversion to a lower frequency means less expensive coax can be used.) We can use Friss' formula again but with one proviso – the received signal has an S/N before it is detected and so the noise that the antenna generates (k*TB*) has to be added in to give the S/N at the input to the receiver. So, taking $B = 30$ MHz and $T = 50$ K as before, the additional thermal noise is 2.1×10^{-14} W. We have an S/N of 6.8×10^3 and a signal power of 6.3 pW. Thus, the total input noise power is

$$\frac{6.3 \times 10^{-12}}{9.7 \times 10^3} + 2.1 \times 10^{-14} = 2.2 \times 10^{-14} \, \text{W}$$

This gives an S/N at the input to the LNB of 286 or 24.6 dB. The noise figure of the LNB is normally quoted and the consumer can specify very low noise LNBs. Some suppliers quote a noise figure of 0.1 dB (1.02 ratio) which is for the whole LNB (Figure 6.2) from the antenna input to the coaxial cable to the set-top box. So, we have an S/N at the input of 286 and therefore the S/N at the output of the LNB is

$$\frac{\text{S}}{\text{N}} = \frac{286}{1.02} = 280 \text{ or } 24.5 \, \text{dB}$$

The LNB unit has to be small enough to fit on an arm of the dish. It must also have enough gain to minimise the effect of the coaxial cable bringing the signal to the set-top box at 1.5 GHz. As for terrestrial TV, orthogonal frequency-division multiplexing (OFDM) and RS error correction are used. (The term *coded OFDM* [COFDM] is often used for this system.)

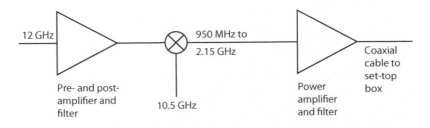

FIGURE 6.2   Block-diagram of a LNB for satellite down-converter.

6.2 ETHERNET

This is the protocol used in "wired" local area networks (LANs). This system supports bit rates of 10 Mbit/s, 100 Mbit/s, 1 Gbit/s, 10 Gbit/s, 100 Gbit/s with 400 Gbit/s being ratified. (There is also talk of a 1 Tbit/s standard.) Speed of transmission is very important and there are two terms in use: *bit rate* – the rate at which data is sent; *baud rate* – the rate at which symbols are sent. As an example, consider 10 Mbit/s Ethernet. The non-return-to-zero (NRZ) data is encoded using Manchester or bi-phase signalling as shown in Figure 6.3. As can be seen, the baud rate of the code is twice the bit rate and this means that the bandwidth requirement is doubled. Manchester coding is also known as two-level pulse position modulation (2-PPM) because the pulse can take on one of two positions within the frame.

As can be seen from Figure 6.3, there are transitions in every bit time and that makes clock extraction very easy – a phase-lock loop (PLL) will suffice. However, this ease of clock extraction has come at the price of a doubling of the bandwidth. This is not so significant at 10 Mbit/s, but is significant when considering higher-speed links. In order to reduce the bandwidth requirement for 100 Mbit/s Ethernet, a three-level signalling format is used – MLT-3 – as well as line coding. Line coding is used to tailor the spectrum of the data to that of the channel. As an example, data can have long sequences of 1s and this causes problems when passing the data through coupling capacitors (baseline wander). The solution is to scramble the data in a line coder.

Figure 6.4 shows the MLT-3 signalling format. When a logic one is sent, the code changes state. A long string of ones would give +1, 0, −1, 0, +1, 0, etc. This staircase waveform obviously has a lower-frequency response than NRZ data. When a zero is coded, the waveform simply maintains its current state. As can be seen from Figure 6.4, the coding of logic ones results in many edges for timing extraction. However, a long run of zeroes will not generate any timing content. The solution is to code the data so that the maximum run of zeroes is limited using a line code. As with so many coding schemes, there is a price to pay in terms of increased line speed. The coding scheme used for 100 Mbit/s Ethernet is 4B5B, i.e. 4 bits coded to 5 bits, resulting in a baud rate of 125 Mbit/s. Appendix XIII lists the coding scheme. As

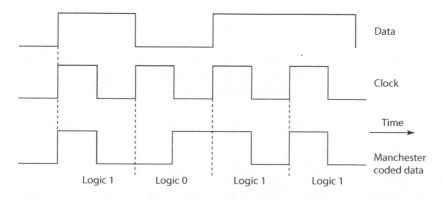

FIGURE 6.3 Production of Manchester encoded data.

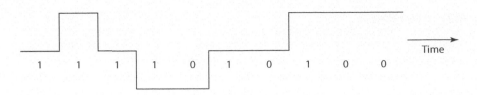

FIGURE 6.4   Generation of the MLT-3 code.

can be seen, the maximum run of logic ones is eight. This does not cause a problem since the logic ones follow a staircase. The number of consecutive logic zeroes is limited to three and this aids timing extraction.

For gigabit Ethernet there is a twisted pair option and an option based on optical fibre. The twisted pair option uses four twisted pairs each carrying 250 Mbit/s. Rather than use NRZ signalling, 1000BASE-T uses a five-level code with trellis code modulation to reduce the line rate. (A discussion of trellis code modulation is beyond the scope of this book but it is similar to convolution coding.) Table 6.2 details the PAM-5 coding scheme. Interface cards can automatically switch bit rates to 100 or 10 Mbit/s.

For the optical option, bandwidth is not such a problem and so the PAM-5 multilevel signalling format is not required. Instead, an 8B10B code is used to provide timing information. The complete code is too long to be reproduced here; however, the code does have zero direct current (dc) content (important to avoid baseline wander) and good timing extraction. There is a 25% overhead in line rate.

For data rates greater than 1 Gbit/s, the line code 64B66B is used. The additional two bits are either 01, in which 64 data bits follows, or 10 in which an 8-bit word is followed by either data or control information. Note that a transition is guaranteed every 66 bits and this aids timing extraction. To further aid timing extraction and alleviate baseline wander, the data is scrambled prior to transmission. This scrambling takes place in a linear feedback shift register. At the receiver the reverse takes place.

In Ethernet, data is divided up into packets which are included in a frame. This frame starts with a seven byte preamble of alternating 1s and 0s which is necessary to synchronise system clocks. The preamble is followed by the one byte start frame delimiter (SFD). This byte signifies the start of the media access code (MAC), which comprises the destination and source code both six bytes long. The data (46–1500 bytes) follows the MAC words and that is followed by four bytes of cyclic redundancy code. A unique stop sequence finishes the frame. Early Ethernet systems used carrier sense multiple access with collision detection (CSMA/CD). This is a simple

TABLE 6.2

PAM-5 Code for 1000BASE-T

Incoming data	000	001	010	011	100	101	110	111
Coded data	0	+1	+2	−1	0	+1	−2	−1

protocol in that a sending station checks the bus (line) to see if the bus is free, i.e. no one is transmitting. If it is free, the station sends its data; if the bus is in use, the station has to wait its turn. Negotiations like this take time and so the actual data rate will be low.

6.3 OPTICAL COMMUNICATIONS

Optical fibre communications have played a major role in the information revolution. Bit rates on trunk route applications are now routinely 100 Gbit/s with 1 Tbit/s links being installed. This is a lot of data, almost unheard of a decade ago. We are now in the situation where we have components that match those of a super-heterodyne radio receiver (mixers and local oscillators) albeit at a far higher frequency. Many wavelengths are currently being used (Table 6.3) although for industrial links 850 nm is widely used. At this wavelength, the sources and detectors for this window are fairly cheap and easy to use and attenuations are of the order of 10 dB/km for this first window. In the bands given in Table 6.3, the typical loss is 0.2 dB/km.

Optical fibre, in general, consists of a high refractive index core surrounded by a lower refractive index cladding (Figure 6.5). The dimension of the cladding is typically fixed at 125 μm so that connectors can be standardised. The dimension of the core ranges from 5 to 75 μm. Light propagates down the core by virtue of total internal reflection and Figure 6.6 shows the total internal reflection at a boundary. When a ray hits the boundary at the critical angle, it runs along the boundary as $\theta_r = 90°$. If the incident angle is greater than the critical, the ray is reflected. This process carries on throughout the length of fibre. A fibre that supports lots of modes (ray paths) is called *multimode fibre* and this is the fibre type most widely used in industrial links. If the core is made sufficiently small (5 μm diameter) then only one ray path can propagate. This is called *single-mode fibre* and it is mainly used in long-haul links.

By applying Snell's law:

$$n_1 \sin \theta_i = n_2 \sin \theta_r \qquad (6.2)$$

TABLE 6.3
Transmission Bands for Optical Communications

Band	Wavelengths (nm)	Comment
O – original	1260–1360	Second transmission window – zero dispersion at 1300 nm
E – extended	1360–1460	
S – short wavelengths	1460–1530	
C – conventional	1530–1565	Third transmission window – very low loss (0.2 dB/km)
L – long wavelengths	1565–1625	
U – ultra long wavelengths	1625–1675	

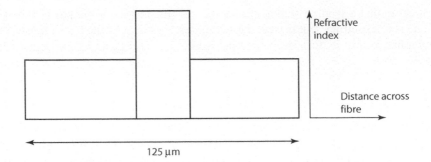

125 μm

FIGURE 6.5   Refractive index profile of optical fibre.

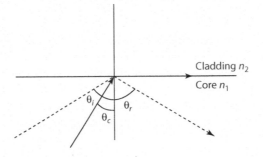

FIGURE 6.6   Total internal reflection between two different materials.

At the critical angle, $\theta_r = 90°$ and so

$$n_1 \sin \theta_i = n_2 \sin 90$$

Therefore,

$$\sin \theta_c = \frac{n_2}{n_1} \tag{6.3}$$

Fibre bandwidth is limited by dispersion – a smearing out of pulses. Consider the cross section of a piece of fibre shown in Figure 6.7. Two ray paths (modes) are drawn; one is straight down the middle of the fibre, while one reflects off the cladding. Both rays carry the signal which, in this instance, is a pulse. The ray that goes straight down the middle will arrive at the destination faster than the ray that bounces off the walls. In addition, because the angled ray travels further, it will be attenuated more than the ray going down the middle. Of course, these are extremes and there are many ray path angles between these limits. (It is not an infinite number as only some modes are allowed.) The net result is a smearing out of the signal and interference between symbols – inter symbol interference (ISI).

For short haul, LAN's large core (50–75 μm) multimode fibre can be used. Indeed, a fibre known as *plastic-coated silica* (PCS) is often used. This has a large core and so lots of light can be coupled into it. For long-haul telecommunications

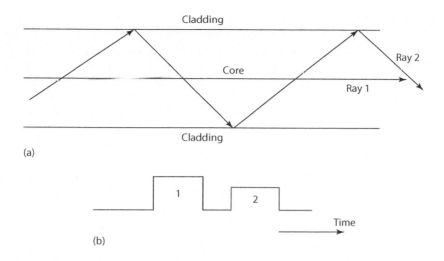

FIGURE 6.7   (a) Maximum and minimum ray paths in fibre and (b) multipath dispersion.

links, single-mode (SM) fibre is used. It might be thought that SM fibre will have an extremely high bandwidth. In principle this is correct, but there are two other forms of dispersion – material and waveguide. Material dispersion comes from different light frequencies travelling at different speeds so that the pulses are smeared out. Fibre has zero material dispersion at 1.3 µm; however, most long-haul links operate at 1.55 µm where there is some negative material dispersion. (A negative dispersion does not mean that the pulse gets smaller as it travels down the fibre! It just means that longer wavelengths travel slower.) Fortunately, waveguide dispersion is positive while material dispersion is negative and it is possible to vary the diameter of the core to balance these two sources of dispersion so that the total dispersion is close to zero.

The transmitter is either a light-emitting diode (LED) or a laser diode. For industrial links and LANs, the choice is an LED for low speeds (<100 Mbit/s) or a vertical-cavity surface-emitting laser (VCSEL) for high speeds. The VCSEL can operate at speeds of 10 Gbit/s, although there are lower speed and cheaper devices. For long-haul routes, lasers are exclusively used. One parameter that is very important is the line width of the source. We are all familiar with the light coming from a red LED and it might be thought that the light is pure. However, that is not the case as the line width of an LED can be as great as 30 nm. To put this in perspective, a wavelength of 650 nm (red) is a frequency of 4.615×10^{14} Hz. This is the nominal centre frequency but it has upper and lower frequencies of 4.724×10^{14} and 4.511×10^{14} Hz. This gives a source with a spread of frequency of 2.13×10^4 GHz. If this were a radio frequency (RF) source, it would take out the whole of the RF spectrum. This spread of light frequencies causes material dispersion but this is usually small when compared to the modal dispersion in multimode fibre. The bandwidth of multimode fibre is of the order of 500 MHz/km and so it can easily be used in small LANs and industrial links.

For high data rate, long-haul applications, a laser operating at a wavelength of 1.55 µm is used. This can be directly modulated by varying the current drive but this is useful up to about 1 Gbit/s with the limit being caused by charge storage in the diode. For higher speeds, an external modulator, a Mach–Zehnder interferometer (MZI), must be used. In this case, the laser is run in continuous wave (CW) mode (it is run continuously without modulation). The interferometer works by splitting the light into two paths, one of which introduces a phase shift. When summed at the output, the light either adds constructively to give light if there is no phase delay, or destructively (no light) if the phase shift is 180°. A schematic of a Mach–Zehnder modulator is shown in Figure 6.8. The waveguides are etched into a crystal of LiNbO$_3$ – lithium niobate – which has the fortunate property that an electric field across the device gives a phase delay.

An alternative to simply changing the amplitude of the light source is differential phase shift keying (DPSK) and Figure 6.9 shows a modulator. In this modulation scheme, the phase remains the same whenever there is a logic 1 regardless of the current state, and changes by 180° whenever a zero is transmitted, again, regardless of the preceding phase. So, arm1 provides the inverted signal and arm2 gives the in-phase signal. Demodulation is really quite easy as all that is required is to add the current symbol with a 1-bit delayed version (Figure 6.9b). Table 6.4 details the

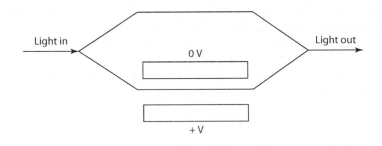

FIGURE 6.8 A schematic of a Mach–Zehnder interferometer.

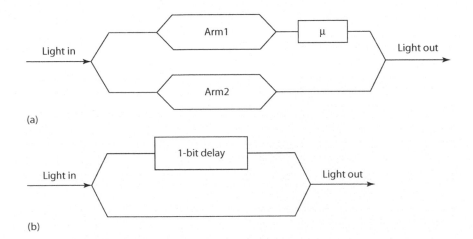

FIGURE 6.9 Schematic of (a) a DPSK modulator and (b) a DSPK demodulator.

TABLE 6.4

Coding Table for DPSK

Original data	0	1	1	1	0	1	0	0	0	1	0
DPSK	0	0	0	0	π	π	0	π	0	0	π
Recovered data		1	1	1	0	1	0	0	0	1	0

coding and decoding of DPSK. The speed of this coding scheme is limited by the technology used to produce the time delay in the receiver. Taking a refractive index of 2.3 for LiNbO$_3$ gives a speed of light of 1.3×10^8 m/s. At a bit rate of 10 Gbit/s, a 180° delay is a length of 13 mm.

There are many different variations to a laser and there are some that have a linewidth measured in kilohertz making them very useful in combating dispersion. It also means that we are able to use coherent detection where the output of a local oscillator is mixed with the received signal. This is identical to the principles we explored previously together with the difficulties – the correct frequency and phase are needed. (See Figure 4.9 and associated text.) It is possible to use an optical PLL, which results in a very sensitive and selective receiver.

The receiver in an optical link is a photodiode (selected for a particular wavelength) followed by a wideband pre-amplifier. For 850 nm links, silicon photodiodes can be used. For longer wavelengths, a mixture of elements is used to give InGaAs photodiodes. Photodiodes, particularly Si, can be quite large and so can collect a lot of light. One thing that all photodiodes suffer with is junction capacitance, which can range from 10 pF down to 0.1 pF. It might appear that the capacitance is very small, but capacitance together with the input resistance of the pre-amplifier gives a time constant that limits the bandwidth. There is also a question of noise in the pre-amplifier and as the signal strength can be very low, this can cause the link to fail.

There are two budgets that concern us in optical communication: the power budget and the dispersion budget. We will consider both of these for short-haul and long-haul links. For the short-haul link, we assume operation at 850 nm and a laser source emitting 5 mW (7 dBm – the "m" in dBm means we are referencing 1 mW). We take a coupling loss of 5 dB, a fibre loss of 10 dB/km and a data rate of 125 Mbit/s. We can draw up a power budget table (Table 6.5) with x being the maximum transmission distance. For the link to operate correctly, the launch power minus the losses must equal the receiver sensitivity (the amount of power needed to get a satisfactory error rate, usually 1 bit in 10^9). So,

$$7 \text{ dBm} - \left(5 \text{ dB} + 10x \text{ dB} + 5 \text{ dB} \right) = -30 \text{ dBm}$$

Therefore, $x = 2.7$ km. This is quite respectable and shows that the link could be used for 100 Mbit/s Ethernet. The operating margin of a few decibels accounts for tolerances in the transmitter and receiver. We should also check on the dispersion over the link. After all, we have a link that should work from a power point of view, but what if the pulses suffer from an excessive amount of ISI?

TABLE 6.5

Power Budget for a Simple Industrial Link

Laser	5 mW = 7 dBm
Coupling loss	5 dB
Fibre loss	10×
Operating margin	5 dB
Receiver sensitivity	−30 dBm

We have chosen a laser as the source and so the linewidth, and hence material dispersion, will be small. As this is an industrial link, we can assume multimode fibre with a modal dispersion of 3 ns per kilometre. Dispersion is related to the maximum bit rate by

$$BR = \frac{0.83}{\pi \times \text{dispersion}} \tag{6.4}$$

Equation 6.4 is an approximation but it does give an idea of how the link will operate. The power budget told us that the link length is 3.2 km and a dispersion of 3 ns/km gives us a total dispersion of 9.6 ns. Using Equation 6.4 gives us 30 Mbit/s. It is clear that the link will not sustain 100 Mbit/s Ethernet. The link is said to be dispersion limited. Of course, if the link length is limited to 500 m, the data rate would be six times greater and would cope with 100 Mbit/s Ethernet.

To turn to the long-haul link, we again use a laser but this time we will use one with a linewidth of 1 MHz. Use of this laser, together with SM fibre, will mean that dispersion effects will be so minimal that they can be ignored for a first pass. We will take a 10 mW, 1550 nm, laser and SM fibre with an attenuation of 0.15 dB/km. We will operate this link at a data rate of 10 Gbit/s and assume a receiver sensitivity of −20 dBm. See Table 6.6 for the power budget table.

So,

$$10 \text{ dBm} - (3 \text{ dB} + 0.15x \text{ dB} + 5 \text{ dB}) = -20 \text{ dBm}$$

which gives $x = 148$ km. What this means is that the pulses can be sent 148 km before they need to be regenerated. A very useful device that is now being used is called

TABLE 6.6

Power Budget for a Telecoms Grade Optical Fibre Link

Laser	10 mW = 10 dBm
Coupling loss	3 dB
Fibre loss	0.15×
Operating margin	5 dB
Receiver sensitivity	−20 dBm

the *erbium-doped fibre amplifier* (EDFA), which amplifies 1550 nm signals so that they do not need electronic regeneration. There are two drawbacks – dispersion is not taken into account and the amplifiers do add some noise (optical noise not electronic). However, this does mean that signals can be sent a very long way without the need to regenerate electronically.

In any communications link there comes a limit on which the speed of a link can operate. This usually comes with the terminal equipment – the logic gates used in detecting the data, the timing extraction circuitry and the optical receiver. At present, 10 Gbit/s links are routine with 100 Gbit/s links being rolled out. These links operate with one transmitter so it is rather like a single radio station and no other. Like the radio station, it is possible to have individual, narrow line width lasers operating at different frequencies. This is known as *wavelength-division multiplexing* (WDM) and is a way of increasing the aggregate bit rate. Indeed, the International Telecommunications Union (ITU) have specified a channel spacing of 50 GHz and this places a restriction on the speed of each data stream (interference could come from the sin x/x spectrum of rectangular pulses). One way of increasing the data rate without increasing the bandwidth is to use two orthogonal states of polarisation. Another is to use quadrature amplitude modulation (QAM) and other multilevel signals. By using such multiplex techniques, an aggregate data rate of 1000 Tbit/s has been achieved. Such speeds come at a cost, but this is an awful lot of data spread over a large number of users so the per user cost can be reasonable.

6.4 MOBILE PHONES

The first-generation mobile phone was analogue in form and was quickly replaced by the digital second-generation (2G) phone. The second generation operated at 900 or 1800 MHz and carried some of the services we use today, albeit at a slow rate (300 kbit/s), but enough for texting. Third-generation (3G) devices provided us with many of the features we have come to expect now. Depending on the type of phone, the data rate can be 0.3–42 Mbit/s, which is sufficient for reading e-mails and video conferencing using Skype; in fact, it is satisfactory for most people's needs. After this there is fourth generation (4G), which operates with a design data rate of 100 Mbit/s. This generation uses Internet protocol (IP) and speech is carried as voice over IP (VoIP). The frequencies used are 850, 900, 1800 and 1900 MHz and some phones are quad-band. There are many evolutions of the standards, so many that they are not listed here. However, it is instructive to review the technology used in the different generations (Table 6.7).

There is major problem with the concept of a mobile phone – what happens if more than one person wishes to make a call? A simple single-frequency system cannot be used as all telephone calls will be carried by one carrier signal. FDM can be used as can time-division multiplexing (TDM). The 2G standard uses both of these, giving 124 channels around 900 MHz. Compression algorithms are used to reduce the bandwidth of each telephone channel. Figure 6.10 shows how users are allocated a time slot in a frame. As we have already seen in Section 3.1, sampling can be used on audio and video provided samples are taken fast enough. If we have a single channel sampled at 8 kHz and 8 bits of data are generated per sample, we have a data rate

TABLE 6.7

Evolution of the Mobile Phone

Generation	Data Rate	Services
1 (1984)	–	Voice communication – analogue
2 (1991)	0.1–0.3 Mbit/s	SMS, MMS and picture messaging
3 (2002)	0.3–42 Mbit/s	Data – Internet, speech
4 (2010)	100 Mbit/s	OFDM Internet protocol–based system – VoIP
5?	1 Gbit/s	All of the above + IoT

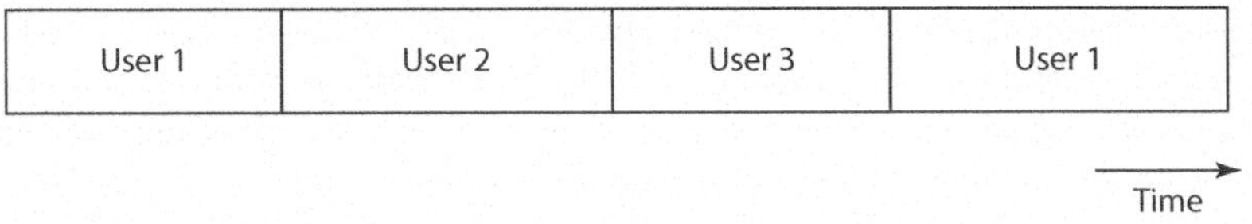

User 1	User 2	User 3	User 1

Time →

FIGURE 6.10   Time-division multiplexing of telephone calls.

of 64 kbit/s. If we want to transmit another data stream, we can reduce the width of the original data stream by a factor of two and interleave the second one in the gap. So, the line rate increases by a factor of two to 128 kbit/s. In Figure 6.10, we have three channels multiplexed and so the line rate is three times the individual data rate. This is known as *time-division multiplexing*. There is a problem with demultiplexing as the clocks at the coder and decoder must be synchronised but this can be done at the start of the call.

An evolution from 2G to 2.5G was the use of packet-switched data (General Packet Radio Service – GPRS). In this system, data is bundled up and attached to a message that has a destination code as its header. The message is reconstructed at the receiver and, if the coding is fast enough, no discernible distortion occurs. FDM and TDM are still used.

3G introduced us to smart phones. The data rate on 3G lies between 144 kbit/s and 2 Mbit/s. This data rate is sufficient to perform video messaging, surf the Internet, e-mail, etc. It does this by using the Universal Mobile Telecommunications System (UMTS), which specifies an uplink frequency range of 1.885–2.025 GHz and a downlink range of 2.11–2.20 GHz. In addition to the large bandwidth, the UMTS also specifies code division multiple access (CDMA) as the modulation format. In this signalling format, the data is gated (XORed) with the output of a pseudo-random binary sequence. At the receiver, use of an identical code can recover the data. The benefit of this system is that many users can use the same frequencies and they will not interfere because they have different codes.

4G is the latest system to be launched. It is designed to have a 100 Mbit/s download speed, although the actual speed is likely to be less. 4G–long-term evolution (LTE) is based around IP and OFDM. As we have already seen, by spreading the data among many carriers, the effective data rate can be increased. In LTE, the

OFDM channels are spaced 15 kHz apart, which equates to 15 ksymbols/second. In a 20 MHz bandwidth, this gives a symbol rate of 18 Msym/s. As we have already seen, we do not need to transmit data as a baseband. Instead, we can use systems such as QAM. Indeed, if 64-QAM is used (representing 6 bits of data) the final data rate is 108 Mbit/s. Remember to place this in context – this data rate is going to a telephone but it is not the audio that demands the high data rate, it is access to the Internet and other services. Work is being done now on fifth generation (5G) but there are difficulties with standardisation. What does appear to be true is that a new standard comes out almost every 10 years. As we approach one phone for every person on the planet, that is an awful lot of data that is invariably carried by optical fibre links as a backbone.

In the United States, there are competing standards with different carriers using different standards – cdmaOne, Global System for Mobile Communications (GSM), CDMA2000, W-CDMA and HSPA+. We have already encountered CDMA in this section and the other CDMA flavours just mentioned are variants on the theme. GSM was the first standard for digital telephones and it specified TDM. HSPA+ stands for evolved high-speed packet access, and it uses multiple in multiple out (MIMO) and 64-QAM to achieve a bit rate of, theoretically, 84.3 Mbit/s. One standard that is agreed on is the use of LTE for 4G.

6.5   DIGITAL AUDIO BROADCASTING

In principle, the conversion of an analogue signal to a digital one is very simple – use an analogue to digital converter (ADC); however, the practice is more difficult. Digital audio broadcasting (DAB) exists in an environment of multipath propagation and so, as discussed in Section 3.3.5, a guard interval is required if a single-frequency network is used. Another point worthy of note is that the S/N for a digital signal falls very rapidly if the carrier power falls, whereas for frequency modulation (FM) the reduction in S/N is not as dramatic. However, the advantage of transmitting digital information is that compression techniques such as Moving Pictures Experts Group (MPEG-2) and error correcting codes can be used. (With the MPEG-2 codec, the bit rate used is typically 128 kbit/s.)

Error correcting codes are used in DAB. The DAB service in use in Europe and the United Kingdom uses a system known as *punctured convolutional coding*. We have already encountered convolutional codes in Section 3.4.4 and the puncturing simply means extracting some of the parity bits, in a predetermined sequence, from a coded data stream. For instance, the puncturing sequence might be 0 1 1 0 0 1 where data is transmitted when there is a 1 and removed when there is a 0. Decoding is by use of the Viterbi decoder. There is a new standard for DAB which is not backwards compatible. The new standard is DAB+ and it uses an audio codec, HE-AACv2, that is more efficient than MPEG-2 giving a bit rate of 64 kbit/s. This reduction in coding leads to a reduction in bit rate and more complex error correction. Indeed, DAB+ uses RS and convolutional coding to generate a low error rate given a poor S/N. OFDM is used to reduce the bit rate per sub-carrier. The bandwidth between each sub-carrier is 1 kHz (corresponding to 1 kbit/s) and this gives a bandwidth of approximately 1.5 MHz.

In the United States, there is a different approach to DAB – HD radio. Here, the digital signal is transmitted alongside the analogue one which could be amplitude modulation (AM) or FM. To keep the bandwidth to a minimum, compression is used along with OFDM. For instance, consider an FM signal transmitting at 100 MHz. Data can be transmitted using OFDM (70 kHz bandwidth) both above and below the FM signal. This, together with a guard band of 60 kHz, gives an overall bandwidth of 400 kHz.

6.6 DIGITAL VIDEO BROADCASTING

Digital TV comes in several different formats – digital video broadcasting-terrestrial (DVB-T), DVB-C (cable), DVB-S (satellite) and 8-vestigial sideband (VSB) in the United States. Europe uses COFDM and QAM up to 256 symbols. This combination allows for a data rate of at least 25 Mbit/s giving high-definition TV channels. (This is quite remarkable given that a high-definition channel could produce data at a rate of 1 Gbit/s.) The audio and video channels are coded using the MPEG-2 algorithm which reduces the redundancy in the signals. (In effect, the things that do not change in an image – the background for instance – are not continually coded. Rather the things that move are.) Systems, either based on OFDM or 8-VSB, start with the same basic blocks.

Different countries have different standards and this complicates the manufacture and sale of TVs and set-top boxes. In general, the bandwidth of a multiplex is 8 MHz, and the number of carriers can be 2k or 8k. Thus, the channel spacing is 4 or 1 kHz. Couple this with high-level QAM (64 or 128) and the aggregate data rate is very large. (High-definition TV requires a higher bit rate than standard definition TV and so it uses QAM 256.) One point worthy of note is that the phase noise of the carrier must be extremely low as the QAM symbols are close together.

In the United States, a system known as 8-VSB is used. This is a multilevel signal with eight levels thus it can carry 3 bits of data. As we have noted when considering AM, there are two sidebands conveying the same information. In VSB, one of the sidebands is filtered but, as real filters are used, a vestige of the sideband remains.

In common with both the COFDM and 8-VSB systems, the input is an MPEG-2 data stream with a data rate of approximately 45 Mbit/s. This data rate is capable of carrying 7 high-definition channels or 25 standard channels. This is quite remarkable when it is noted that the data rate entering the MPEG-2 encoder could be as high as 1 Gbit/s. These figures clearly show the redundancy inherent in TV pictures. In the RS encoder, extra parity bytes are added to the MPEG-2 data stream. The precise number is a variable set by the operator of the codec. Convolutional coding is used after RS encoding. It is worth remembering that this is all done at a relatively low speed and so the overhead, in terms of speed, associated with the processing is not great. Ultimately, what we are left with is a multiplex, either 8-VSB or OFDM, at a low frequency. Multiplication stages can be used to translate the signal to its final broadcast frequency.

6.7 WI-FI

There are many standards for Wi-Fi but their common base number is 802.11 and they operate at frequencies of 2.4 and 5 GHz. Rather than examine all the possible variants, we will only look at 802.11n and the future trends.

IEEE 802.11n was first ratified in 2009. It is a dual-band system operating at 2.4 or 5 GHz. (It can use 2.4 GHz provided it does not interfere with other users, or has interference itself. The specified bandwidth is 2.412–2.472 GHz.) The standard also specifies for MIMO technology, which is a form of spatial multiplexing (discussed next). We can estimate data throughput by considering a 20 MHz wide channel with a fast Fourier transform (FFT) of 64 giving 52 OFDM data channels and 4 pilot tones. The channel spacing is 20 MHz/64 = 312.5 kHz. Each of the carriers can be modulated up to 64 QAM giving a total bit rate of approximately 90 Mbit/s. Four separate MIMO channels would give a data rate of 360 Mbit/s. (There is an overhead for error correction as well as the need for a guard interval at the end of each symbol.)

Later standards such as 802.11ac have a bandwidth of 160 MHz in the 5 GHz band to deliver a data rate of nearly 900 Mbit/s. It does this using MIMO technology and OFDM carriers. Still greater bit rates are available if the 802.11ad standard is used. This operates at 60 GHz to give a data rate of 6.75 Gbit/s. We should always be conscious of the data rate that these services require. The high data rates for Wi-Fi have to come from a backbone transmission system. If the system is optical fibre into the home, there will be sufficient bandwidth, but if the line is copper there will be a bottleneck and transmission rates will be limited.

6.8   MIMO

MIMO consists of multiple transmit antennae and multiple receive stations. In effect, it is spatial diversity in that the same signal can be transmitted on all transmit antennae or different data streams can be transmitted on different antennae and decoded at the receiver. To explain, let us consider single in single out (SISO). Here, we have one transmitter and one receiver. This is basically a single channel radio system. There are problems with this system: it is susceptible to multipath distortion, selective fading and loss of signal due to obstacles. An alternative is to use single in multiple out (SIMO). The multiple receive antennae detect the transmit signal via several different paths and the signals arrive at different times. It may be that the time difference is very small (of the order of nanoseconds), but it can be similar to the bit time we are transmitting. (A 100 Mbit/s data stream has a bit time of 10 ns.) At the receiver, the data streams are added together by introducing time delays to compensate for the multipath effects. MIMO is simply multiple transmit antennae and multiple receive antennae. There can be multiple channels carrying separate data, or a single channel of data. This is the system used in 802.11ac and it can deliver very high data rates. A necessary innovation is to characterise each channel in terms of attenuation and time delay (channel state information – CSI) so that the data can be recovered correctly.

6.9   ASYMMETRIC DIGITAL SUBSCRIBER LINE

Asymmetric digital subscriber line (ADSL) is the means by which most of us without optical fibre or satellite modems receive digital data. It operates over standard telephone lines by using frequencies that are above the speech signals (restricted to 3.4 kHz). There are two bands in operation: 25.875–138 kHz for the uplink and

138–1104 kHz for the downlink. The uplink is slower (3.3 Mbit/s) than the downlink (24 Mbit/s) because of the smaller bandwidth, but this does not cause too much of a problem because downloading data (films and music) is a major part of Internet use.

There are many different signalling protocols in use in ADSL: two-binary, one-quaternary (2B1Q); carrierless amplitude phase (CAP) modulation; trellis-coded pulse-amplitude modulation (TC-PAM); discrete multitone modulation (DMT); or OFDM. Of these four, OFDM with QAM is the most widely used as it gives a very high aggregate data rate. Filters are required to split off the data stream from the audio signal. In view of the frequencies involved, there is the risk of interference from AM transmissions.

6.10 BLUETOOTH

Most of us are familiar with Bluetooth as a means to connect our mobile phones to the audio system in our cars. In common with Wi-Fi, the Bluetooth standard has gone through many revisions – it is now up to version 5. Currently, Bluetooth uses the 2.4 GHz band with a range of frequencies from 2.402 to 2.480 GHz split into 79 channels with a 1 MHz spacing. Rather than transmit discrete data on each of the channels, like OFDM, data is spread over the channels using a pseudo-random sequence. This is known as *frequency hopping* and Bluetooth uses 800 hops per second. In order to decode the data correctly, the receiver has to synchronise to the transmitter and follow the hops. It does this by using the same pseudo-random sequence as the transmitter. This pseudo-random spread of data means that the signal appears to be wideband noise meaning that other users with narrowband signals are not subject to interference.

The data rate for version 5.0 is 50 Mbit/s under ideal conditions. This may not seem particularly high but is sufficient for connecting computer peripherals and forming a network. A possible scenario would be a Bluetooth network operating at a low data rate with a bridge to a Wi-Fi network either via a computer or directly to a Wi-Fi hub.

6.11 THE INTELLIGENT HOME

We have spent a great deal of time discussing high-speed links such as Wi-Fi and optical fibre; however, in the intelligent home we do not need high-speed links or even long-haul links. Instead, we require local networks that can be as local as a room. There are two systems available two us: Zigbee and power line communications (PLC). At first it might be thought that a Zigbee network and a Bluetooth network fulfil the same purpose – they are both RF networks and they can be used in an ad hoc way. To some extent this is true; however, their applications are totally different. Zigbee units are designed to be powered from batteries for at least two years; they operate with low RF power; their range is limited; and their data rate is low. These properties are put to good use if Zigbee is used in a personal area network (PAN). In such networks, for example the Internet of Things (IoT), the data rate is low (a room thermostat does not need full, high-definition TV) and battery life needs to be long (>2 years). Zigbee fulfils these requirements – 65 560 units in a network; the number of hops (between nodes) is 30; battery life is >2 years by putting the end units to sleep for a time; transmission distance is 10–100 m. Zigbee uses several different frequencies depending on where it is operating – 784 MHz in China, 868 MHz in Europe, 915 MHz in Australia and the United States, 2.4 GHz everywhere. The 2.4 GHz option gives a data rate of 250 kbit/s with 16-ary orthogonal *quadrature phase shift keying* (QPSK), while the lowest frequency band gives a data rate of 20 kbit/s with binary PSK. These are quite modest bit rates but it should be remembered that a Zigbee network is likely to be used in intelligent houses and buildings where a high data rate is not required to operate light switches and thermostats. There are 10 channels in the 915 MHz band that start at 906 MHz with a channel spacing of 2 MHz. These are Zigbee channels 1–10. Channels 11–26 occupy frequencies of 2.405–2.480 GHz with a channel spacing of 5 MHz.

A Zigbee network consists of one co-ordinator, several routers and numerous end devices. The last are run off batteries and, in view of this, they are not always on, unlike the co-ordinator and routers that run off the mains. Figure 6.11 shows a typical network as might be found in a two-storey, two-bedroom, one-bathroom house. There are two rooms downstairs and three rooms upstairs. Each room has a router associated with it and each router has a number of end devices.

It may be thought to be a bit of overkill having intelligent lights and room thermostats in every room. However, independent heating makes sense and having control over lighting is useful from a security point of view in that lights can be programmed in sequence to simulate someone going to the bathroom at night. One major advantage of an RF network is that there are no wires (by definition) and so installation is easier than a wired system. It does rely on having Zigbee compatible peripherals, but these are coming to market.

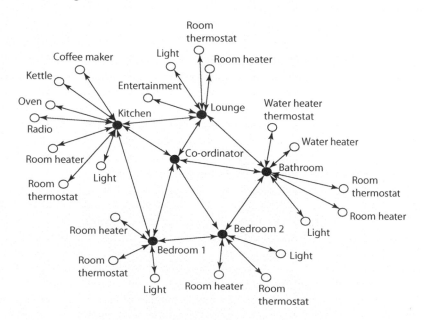

FIGURE 6.11   An example of a mesh Zigbee system.

Zigbee and Bluetooth are RF-based networks. An alternative is Power Line Communication (PLC), which can be used for home automation as well as control over a larger area. The advantage of PLC should be evident – it is carried by mains wiring which goes to just about everything in the house so that there is no need for radio transceivers or additional wiring. PLC uses two frequency bands: 3–500 kHz (narrowband) and 1.8–250 MHz (wideband). The data rates are typically 100 kbit/s for the narrowband and several hundred megabits per second for the wideband link. Data is modulated onto a carrier using QAM techniques, together with RS coding, prior to coupling onto the power line. Unfortunately, there is nothing to stop a house's carrier being picked up by the neighbours. To stop this happening, each house has a unique address. A further problem is that the wiring acts like an antenna, both receiving and transmitting, and so there is the potential for interference. In spite of this, PLC is gaining in popularity for Internet traffic and house control. Narrowband PLC is also very useful for control purposes outside the house. It can be used to turn on and off street lamps and also tell whether the lamps need replacing.

There is a problem with the human interface. With all these sensors and nodes, how are we to control them? An app on our mobile phone would be quite complex if we are to control individual lights in the house. A possible solution is to make the house intelligent with the application of artificial intelligence (AI). The house could know who is in which room and adjust the lighting and temperature according to the occupant. It would do this after learning the preferences of each inhabitant. The house would adjust to each person without the need for many remote controls or a complex app.

6.12   SOFTWARE-DEFINED RADIO

We met the traditional superheterodyne receiver in Section 1.7 where the received signal was mixed to an intermediate frequency (IF) prior to demodulation. With the advent of ultra-fast ADCs, which operate at 3.6 GSample/s, conversion of RF directly to digital is possible. Once it is in digital form, it can be filtered by a finite impulse response (FIR) filter to select a particular station or signal. Decoding of forward error correction (FEC) can then be applied prior to reconstruction in a digital to analogue converter (DAC) to get the modulation back. By doing this, it is possible to remove mixing to an IF completely, thereby simplifying the receiver design. The reverse takes place in the transmitter – the modulating signal is digitised in an ADC, FEC is added and a DAC converts the signal for transmission.

The advantages of software-defined radio (SDR) are that FIR filters are used instead of tuned circuits and they can have a linear phase response and sharp cutoff; the receiver or transmitter can be reconfigured easily; there is no need to align stages in frequency; different standards are accommodated for by simply altering the software; a single transmitter or receiver can be used for various modulation formats with changes being made by software. SDR is certain to have an impact on telecommunications in the future. One receiver on the market covers 1 kHz to 2 GHz without the aid of mixers. It does have a number of hardware filters to select a particular band and the antenna has to be changed as well, but this is a small price to pay for flexibility.

Appendix I: The Double Balanced Mixer

Double balanced mixers come in two types – active (ones that have gain) and passive (ones that have loss). Of the active types, the LM1496 is popular for operation at less than 10 MHz. For higher-frequency operation, the diode mixer of Figure AI.1 can be used even at very high frequencies – 40 GHz.

If the carrier amplitude is high enough, the diodes will switch according to the polarity of the carrier. In the positive half of the cycle, diodes D1 and D2 will conduct and D3 and D4 will be reverse biased and so the modulation is connected to the output. In the negative half cycle, the situation is reversed with D1 and D2 reverse biased and D3 and D4 forward biased. This will have the effect of connecting the inverse of the modulation to the output. Recall that the carrier frequency is higher than the modulation and so the modulation is chopped up by the carrier signal to give positive and negative values (Figure AI.2). What we have produced is double

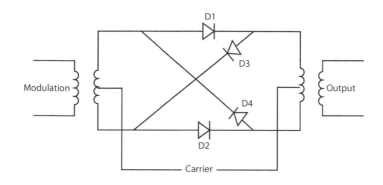

FIGURE AI.1   The double balanced mixer.

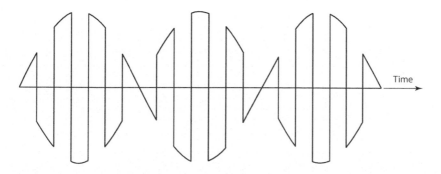

FIGURE AI.2   Output of the double balanced mixer.

sideband suppressed carrier (DSB-SC) transmission, i.e. the modulation and the carrier have been multiplied together. It should be noted that the carrier signal is either on or off – it appears to be a square wave. This will produce harmonics which have to be filtered out. (This is not a problem because we get sum and difference components out of the modulator and we have to filter out the unwanted component.)

Appendix II: The Product of Two Cosines

The product of two cosines can be found from

$$\cos(A+B) = \cos(A)\cos(B) - \sin(A)\sin(B) \qquad \text{(AII.1)}$$

$$\cos(A-B) = \cos(A)\cos(B) + \sin(A)\sin(B) \qquad \text{(AII.2)}$$

Adding Equations AII.1 and AII.2 gives

$$2\cos(A)\cos(B) = \cos(A+B) + \cos(A-B)$$

$$\cos(A)\cos(B) = \frac{1}{2}\cos(A+B) + \frac{1}{2}\cos(A-B) \qquad \text{(AII.3)}$$

Appendix III: The Parallel Tuned Circuit

Figure AIII.1a shows a parallel tuned circuit with parasitic series resistance in the inductor and Figure AIII.1b shows how the reactance varies and gives the resonant frequency. The impedance of the circuit is found from

$$Z = \frac{(R + j\omega L)\,{}^{1}\!/_{j\omega C}}{R + j\omega L + {}^{1}\!/_{j\omega C}}$$

$$= \frac{(R + j\omega L)\,{}^{1}\!/_{j\omega C}}{R + j\left(\omega L - {}^{1}\!/_{\omega C}\right)}$$

$$= \frac{(R + j\omega L)\,{}^{1}\!/_{j\omega C}}{R + j\omega L\left(1 - {}^{1}\!/_{\omega^{2}LC}\right)} \tag{AIII.1}$$

$$= \frac{R\left(1 + {}^{j\omega L}\!/_{R}\right){}^{1}\!/_{j\omega C}}{R(1 + {}^{j\omega L}\!/_{R}\left(1 - {}^{1}\!/_{\omega^{2}LC}\right)}$$

$$= \frac{{}^{L}\!/_{RC} + {}^{1}\!/_{j\omega C}}{1 + {}^{j\omega L}\!/_{R}\left(1 - {}^{1}\!/_{\omega^{2}LC}\right)}$$

At resonance, the impedance is maximum for the parallel tuned circuit. This occurs when the denominator in Equation AIII.1 is minimum, i.e. when $1 - {}^{1}\!/_{\omega^{2}LC} = 0$. Thus,

$$\omega_{o} = {}^{1}\!/_{\sqrt{LC}} \tag{AIII.2}$$

Also note that the impedance at resonance (the dynamic impedance) is

$$Z_{d} = {}^{L}\!/_{RC} + {}^{1}\!/_{j\omega C} \tag{AIII.3}$$

The Q of the circuit is defined as

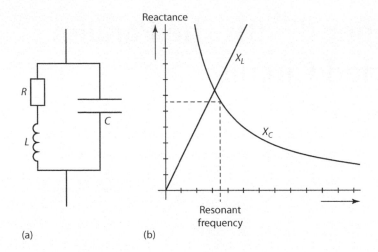

FIGURE AIII.1   (a) A parallel tuned circuit and (b) a plot of X_L and X_C with ω.

$$Q = 2\pi \frac{\text{energy stored}}{\text{energy lost per cycle}}$$

$$= 2\pi \frac{\text{energy stored}}{\text{power lost times } T}$$

$$= 2\pi f_o \frac{\text{energy stored}}{\text{power lost}} \qquad \text{(AIII.4)}$$

$$= \omega_o \frac{L I\text{rms}^2}{I\text{rms}^2 R}$$

$$= \frac{\omega_o L}{R}$$

$$= \frac{f_o}{\Delta f} \qquad \text{(AIII.5)}$$

The current through the inductor and capacitor can be found as follows:

$$I_{\text{supply}} = I_L + I_C$$

$$= \frac{V}{R + j\omega L} + V j\omega C \qquad \text{(AIII.6)}$$

$$= \frac{V \,/ \tan^{-1}\left(-\omega L\big/R\right)}{\sqrt{R^2 + \omega^2 L^2}} + V\omega C \,/\, 90°$$

At resonance, Equation AIII.6 becomes

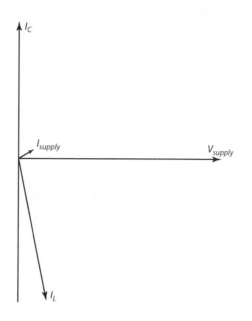

FIGURE AIII.2   A Phasor diagram for a tuned circuit.

$$\frac{V\sqrt{\dfrac{C}{L}} \bigg/ \tan^{-1}\left(-\dfrac{1}{R}\sqrt{\dfrac{L}{C}}\right)}{\sqrt{R^{2(C/L)}+1}} + V\sqrt{\dfrac{C}{L}} \bigg/ 90° \qquad\qquad \text{(AIII.7)}$$

If the series resistance R is zero – the ideal case – the inductor and capacitor currents are identical but 180° out of phase. Thus, they cancel out, the supply current is zero and this gives an infinite magnification factor, Q. If the resistance is non-zero, the currents will not cancel out, Q will be finite and there will be a phase shift associated with the inductive current as shown in Figure AIII.2.

Examples of Q values are 10 and 100. There is an interplay between resonance and Q – a high resonant frequency gives a low Q.

Appendix IV: Decibels

The definition of decibel (dB) is

$$dB = 10\log\left(\frac{Power_{out}}{Power_{in}}\right) \qquad\qquad (AIV.1)$$

Any power ratio can be expressed in decibels. If the output power is less than the input power (as in an attenuator) the decibel will be negative.

It is also possible to use a voltage ratio as follows:

$$Power_{out} = \frac{V_{out}{}^2}{R}$$

and

$$Power_{in} = \frac{V_{in}{}^2}{R}$$

Thus, if the resistance is the same

$$\frac{Power_{out}}{Power_{in}} = \frac{V_{out}{}^2}{V_{in}{}^2}$$

and so

$$dB = 10\log\left(\frac{V_{out}{}^2}{V_{in}{}^2}\right)$$

$$= 10\log\left(\frac{V_{out}}{V_{in}}\right)^2 \qquad\qquad (AIV.2)$$

$$= 20\log\frac{V_{out}}{V_{in}}$$

Appendix V: Noise Factor and Friss' Formula

The noise factor of a network (an amplifier for instance) is given by

$$F = \frac{S_i/N_i}{S_o/N_o} \qquad \text{(AV.1)}$$

where S_i/N_i is the input signal to noise ratio and S_o/N_o is the output signal to noise ratio. Also, the output signal power is

$$S_o = GS_i$$

The output noise comprises the internal amplifier noise, N_a, and the amplified input noise:

$$N_o = N_a + GN_i$$

where G is the power gain of the amplifier. Thus, $F1$ becomes,

$$F = \frac{S_i/N_i}{GS_i/N_a + GN_i} \qquad \text{(AV.2)}$$

$$= \frac{N_a + GN_i}{GN_i}$$

The input noise for a matched system is kTB and so Equation AV.2 becomes

$$F = \frac{N_a + GkTB}{GkTB} \qquad \text{(AV.3)}$$

$$N_a = (F-1)GkTB \qquad \text{(AV.4)}$$

For an attenuator (cable for instance) the gain is less than 1. It is more usual to refer to the loss of the network, L, which is equal to the input over the output or 1/gain. The noise from the attenuator, N_{loss}, is

$$N_{\text{loss}} = N_i - N_o \qquad \text{(AV.5)}$$

The noise generated by the lossy component is equal to the difference between the input noise (which is big) and the output noise, which is smaller due to the loss. Rearranging Equation AV.5 gives

$$N_{\text{loss}} = N_i \left(1 - \frac{N_o}{N_i}\right)$$

$$= N_i \left(1 - \frac{1}{L}\right)$$

$$= kTB \left(1 - \frac{1}{L}\right)$$

Equating to Equation AV.4 gives

$$(F-1)GkTB = kTB \left(1 - \frac{1}{L}\right)$$

Noting that $G = 1/L$

$$(F-1)\frac{1}{L}kTB = kTB \left(1 - \frac{1}{L}\right)$$

giving, after some simplification,

$$F_{\text{loss}} = L \tag{AV.6}$$

Considering a cascade of two stages, the noise at the output of stage one, N_{o1}, is made up of the input noise multiplied by the power gain added to the amplifier noise (Equation AV.4). So,

$$N_{o1} = kTBG_1 + (F_1 - 1)kTBG_1$$

$$= F_1 G_1 kTB \tag{AV.7}$$

This noise forms the input noise to the second stage. Thus, the noise at the output of the second stage is

$$N_{o2} = G_2 N_{o1} + kTBG_2 (F_2 - 1)$$

$$= kTB\{G_1 G_2 F_1 + G_2 F_2 - G_2\}$$

$$= kTBG_1 G_2 \left\{F_1 + \frac{F_2 - 1}{G_1}\right\} \tag{AV.8}$$

This can be compared to the general form of Equation AV.7 and so the gain is the product of the individual gains and the noise factor is

$$F_{\text{total}} = 1 + (F_1 - 1) + \frac{(F_2 - 1)}{G_1} + \frac{(F_3 - 1)}{G_2 G_1} + \cdots \tag{AV.9}$$

Appendix VI: Maximum Power Transfer

As shown in Figure AVI.1, let a source of emf E volts and internal resistance R_{source} ohm be connected to a load of resistance R_{load} ohm. There is a potential divider at work here and so the voltage across the load resistor is

$$V_L = \frac{E}{R_{\text{source}} + R_{\text{load}}} . R_{\text{load}} \qquad \text{(AVI.1)}$$

Giving a current through the load of

$$I_L = \frac{E}{R_{\text{source}} + R_{\text{load}}} \qquad \text{(AVI.2)}$$

The power is the product of the voltage and the current and so

$$
\begin{aligned}
\text{Power} &= \left(\frac{E}{R_{\text{source}} + R_{\text{load}}} \right)^2 . R_{\text{load}} \\[2mm]
&= \frac{E^2}{R_{\text{source}}^2 + 2 R_{\text{source}} R_{\text{load}} + R_{\text{load}}^2} . R_{\text{load}} \qquad \text{(AVI.3)} \\[2mm]
&= \frac{E^2}{R_{\text{source}}^2 \big/ R_{\text{load}} + 2 R_{\text{source}} + R_{\text{load}}}
\end{aligned}
$$

This power will be at a maximum when the denominator in Equation AVI.3 is at a minimum. To find this, the denominator is differentiated by R_{load} and equated to zero to give

$$-R_{\text{source}}^2 \big/ R_{\text{load}}^2 + 1 = 0$$

$$R_{\text{source}} \big/ R_{\text{load}} = \pm 1$$

As resistance is non-negative:

$$R_{\text{load}} = R_{\text{source}} \qquad \text{(AVI.4)}$$

Differentiation of the denominator in Equation AVI.3 twice gives

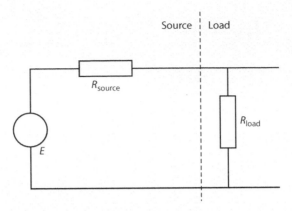

FIGURE AVI.1 Source and load resistance.

$$2\,{R_{\text{source}}}^2 \Big/ {R_{\text{load}}}^3$$

As this is positive, the condition for maximum power transfer is that source and load should have the same resistance. Thus, if the source is an antenna with a terminal resistance of 50 Ω, the load resistance should also be 50 Ω.

If the source has a reactive component, the following (Figure AVI.2) applies:

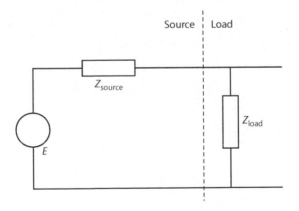

FIGURE AVI.2 Source and load impedance.

where $Z_{\text{source}} = R_{\text{source}} + X_{\text{source}}$ and $Z_{\text{load}} = R_{\text{load}} + X_{\text{load}}$ and X denotes reactance either capacitive or inductive. The phasor current is given by

$$|I| = \frac{|E|}{|Z_{\text{source}} + Z_{\text{load}}|}$$

This current generates an average power in the load of $I_{rms}^2 \cdot R_{load}$. Substituting in for the current gives

$$P_{load} = \frac{1}{2} \frac{|E|^2}{|Z_{source} + Z_{load}|^2}$$

$$= \frac{1}{2} \frac{|E|^2}{(R_{source} + R_{load})^2 + (X_{source} + X_{load})^2}$$

(AVI.5)

The maximum power occurs when the denominator in Equation AVI.5 reaches a minimum. As reactance can be either positive (inductive) or negative (capacitive), it is possible to remove the reactance completely, i.e. $X_{source} = X_{load}$ but of opposite polarity. (If the source is inductive, the load reactance should be capacitive and vice versa.) This is conjugate matching. The maximum power condition is the same as for the resistive condition, i.e. $R_{load} = R_{source}$. So, the matching condition is $Z_{load} = Z_{source}^*$ where the star indicates complex conjugate.

Appendix VII: Error Function (erf) Tables

x	$\text{erf}(x)$	x	$\text{erf}(x)$	x	$\text{erf}(x)$	x	$\text{erf}(x)$
0.00	0.5000	1.00	0.1587	2.00	0.0228	3.00	0.99997791
0.01	0.4960	1.01	0.1562	2.01	0.0222	3.01	0.99997926
0.02	0.4920	1.02	0.1539	2.02	0.0217	3.02	0.99998053
0.03	0.4880	1.03	0.1515	2.03	0.0212	3.03	0.99998173
0.04	0.4840	1.04	0.1492	2.04	0.0207	3.04	0.99998286
0.05	0.4801	1.05	0.1469	2.05	0.0202	3.05	0.99998392
0.06	0.4761	1.06	0.1446	2.06	0.0197	3.06	0.99998492
0.07	0.4721	1.07	0.1423	2.07	0.0192	3.07	0.99998586
0.08	0.4681	1.08	0.1401	2.08	0.0188	3.08	0.99998674
0.09	0.4641	1.09	0.1379	2.09	0.0183	3.09	0.99998757
0.10	0.4602	1.10	0.1357	2.10	0.0179	3.10	0.99998835
0.11	0.4562	1.11	0.1335	2.11	0.0174	3.11	0.99998908
0.12	0.4522	1.12	0.1314	2.12	0.0170	3.12	0.99998977
0.13	0.4483	1.13	0.1292	2.13	0.0166	3.13	0.99999042
0.14	0.4443	1.14	0.1271	2.14	0.0162	3.14	0.99999103
0.15	0.4404	1.15	0.1251	2.15	0.0158	3.15	0.99999160
0.16	0.4364	1.16	0.1230	2.16	0.0154	3.16	0.99999214
0.17	0.4325	1.17	0.1210	2.17	0.0150	3.17	0.99999264
0.18	0.4286	1.18	0.1190	2.18	0.0146	3.18	0.99999311
0.19	0.4247	1.19	0.1170	2.19	0.0143	3.19	0.99999356
0.20	0.4207	1.20	0.1151	2.20	0.0139	3.20	0.99999397
0.21	0.4168	1.21	0.1131	2.21	0.0136	3.21	0.99999436
0.22	0.4129	1.22	0.1112	2.22	0.0132	3.22	0.99999473
0.23	0.4090	1.23	0.1093	2.23	0.0129	3.23	0.99999507
0.24	0.4052	1.24	0.1075	2.24	0.0125	3.24	0.99999540
0.25	0.4013	1.25	0.1056	2.25	0.0122	3.25	0.99999570
0.26	0.3974	1.26	0.1038	2.26	0.0119	3.26	0.99999598
0.27	0.3936	1.27	0.1020	2.27	0.0116	3.27	0.99999624
0.28	0.3897	1.28	0.1003	2.28	0.0113	3.28	0.99999649
0.29	0.3859	1.29	0.0985	2.29	0.0110	3.29	0.99999672
0.30	0.3821	1.30	0.0968	2.30	0.0107	3.30	0.99999694
0.31	0.3783	1.31	0.0951	2.31	0.0104	3.31	0.99999715
0.32	0.3745	1.32	0.0934	2.32	0.0102	3.32	0.99999734
0.33	0.3707	1.33	0.0918	2.33	0.00990	3.33	0.99999751
0.34	0.3669	1.34	0.0901	2.34	0.00964	3.34	0.99999768
0.35	0.3632	1.35	0.0885	2.35	0.00939	3.35	0.99999784
0.36	0.3594	1.36	0.0869	2.36	0.00914	3.36	0.99999798
0.37	0.3557	1.37	0.0853	2.37	0.00889	3.37	0.99999812

(Continued)

(CONTINUED)

x	erf(x)	x	erf(x)	x	erf(x)	x	erf(x)
0.38	0.3520	1.38	0.0838	2.38	0.00866	3.38	0.99999825
0.39	0.3483	1.39	0.0823	2.39	0.00842	3.39	0.99999837
0.40	0.3446	1.40	0.0808	2.40	0.00820	3.40	0.99999848
0.41	0.3409	1.41	0.0793	2.41	0.00798	3.41	0.99999858
0.42	0.3372	1.42	0.0778	2.42	0.00776	3.42	0.99999868
0.43	0.3336	1.43	0.0764	2.43	0.00755	3.43	0.99999877
0.44	0.3300	1.44	0.0749	2.44	0.00734	3.44	0.99999886
0.45	0.3264	1.45	0.0735	2.45	0.00714	3.45	0.99999893
0.46	0.3228	1.46	0.0721	2.46	0.00695	3.46	0.99999901
0.47	0.3192	1.47	0.0708	2.47	0.00676	3.47	0.999999077
0.48	0.3156	1.48	0.0694	2.48	0.00657	3.48	0.999999141
0.49	0.3121	1.49	0.0681	2.49	0.00639	3.49	0.999999201
0.50	0.3085	1.50	0.0668	2.50	0.00621	3.50	0.999999257
0.51	0.3050	1.51	0.0655	2.51	0.00604	3.51	0.999999309
0.52	0.3015	1.52	0.0643	2.52	0.00587	3.52	0.999999358
0.53	0.2981	1.53	0.0630	2.53	0.00570	3.53	0.999999403
0.54	0.2946	1.54	0.0618	2.54	0.00554	3.54	0.999999445
0.55	0.2612	1.55	0.0606	2.55	0.00539	3.55	0.999999485
0.56	0.2877	1.56	0.0594	2.56	0.00523	3.56	0.999999521
0.57	0.2843	1.57	0.0582	2.57	0.00508	3.57	0.999999555
0.58	0.2810	1.58	0.0571	2.58	0.00494	3.58	0.999999587
0.59	0.2776	1.59	0.0559	2.59	0.00480	3.59	0.999999617
0.60	0.2743	1.60	0.0548	2.60	0.00466	3.60	0.999999644
0.61	0.2709	1.61	0.0537	2.61	0.00453	3.61	0.999999670
0.62	0.2676	1.62	0.0526	2.62	0.00440	3.62	0.999999694
0.63	0.2643	1.63	0.0516	2.63	0.00427	3.63	0.999999716
0.64	0.2611	1.64	0.0505	2.64	0.00415	3.64	0.999999736
0.65	0.2578	1.65	0.0495	2.65	0.00402	3.65	0.999999756
0.66	0.2546	1.66	0.0485	2.66	0.00391	3.66	0.999999773
0.67	0.2514	1.67	0.0475	2.67	0.00379	3.67	0.999999790
0.68	0.2483	1.68	0.0465	2.68	0.00368	3.68	0.999999805
0.69	0.2451	1.69	0.0455	2.69	0.00357	3.69	0.999999820
0.70	0.2420	1.70	0.0446	2.70	0.00347	3.70	0.999999833
0.71	0.2389	1.71	0.0436	2.71	0.00336	3.71	0.999999845
0.72	0.2358	1.72	0.0427	2.72	0.00326	3.72	0.999999857
0.73	0.2327	1.73	0.0418	2.73	0.00317	3.73	0.999999867
0.74	0.2296	1.74	0.0409	2.74	0.00307	3.74	0.999999877
0.75	0.2266	1.75	0.0401	2.75	0.00298	3.75	0.999999886
0.76	0.2236	1.76	0.0392	2.76	0.00289	3.76	0.999999895
0.77	0.2206	1.77	0.0384	2.77	0.00280	3.77	0.999999903
0.78	0.2177	1.78	0.0375	2.78	0.00272	3.78	0.999999910
0.79	0.2148	1.79	0.0367	2.79	0.00264	3.79	0.999999917
0.80	0.2119	1.80	0.0359	2.80	0.00256	3.80	0.999999923
0.81	0.2090	1.81	0.0351	2.81	0.999929	3.81	0.999999929

(Continued)

(CONTINUED)

x	$\mathbf{erf}(x)$	x	$\mathbf{erf}(x)$	x	$\mathbf{erf}(x)$	x	$\mathbf{erf}(x)$
0.82	0.2061	1.82	0.0344	2.82	0.999933	3.82	0.999999934
0.83	0.2033	1.83	0.0336	2.83	0.999937	3.83	0.999999939
0.84	0.2005	1.84	0.0329	2.84	0.999941	3.84	0.999999944
0.85	0.1977	1.85	0.0322	2.85	0.999944	3.85	0.999999948
0.86	0.1949	1.86	0.0314	2.86	0.999948	3.86	0.999999952
0.87	0.1922	1.87	0.0307	2.87	0.999951	3.87	0.999999956
0.88	0.1894	1.88	0.0301	2.88	0.999954	3.88	0.999999959
0.89	0.1867	1.89	0.0294	2.89	0.999956	3.89	0.999999962
0.90	0.1841	1.90	0.0287	2.90	0.999959	3.90	0.999999965
0.91	0.1814	1.91	0.0281	2.91	0.999961	3.91	0.999999968
0.92	0.1788	1.92	0.0274	2.92	0.999964	3.92	0.999999970
0.93	0.1762	1.93	0.0268	2.93	0.999966	3.93	0.999999973
0.94	0.1736	1.94	0.0262	2.94	0.999968	3.94	0.999999975
0.95	0.1711	1.95	0.0256	2.95	0.999970	3.95	0.999999977
0.96	0.1685	1.96	0.0250	2.96	0.999972	3.96	0.999999979
0.97	0.1660	1.97	0.0244	2.97	0.999973	3.97	0.999999980
0.98	0.1635	1.98	0.0239	2.98	0.999975	3.98	0.999999982
0.99	0.1611	1.99	0.0233	2.99	0.999977	3.99	0.999999983

Appendix VIII: The Discrete Fourier Transform

The fast Fourier transform (FFT) and inverse fast Fourier transform (IFFT) are closely related to the discrete Fourier transform (DFT) and inverse DFT. Consider a sine wave that is sampled eight times in its period – the sampling rate has to be high enough to satisfy the Nyquist sampling criteria, which means that we must sample at least twice within a cycle. The sine wave is shown in Figure AVIII.1.

As shown in Figure AVIII.1, eight discrete samples are taken in one cycle ($0 \leq n \leq 7$). Taking a period of 1 ms, the frequency of the sine wave is 1 kHz and the sampling frequency is 8 kHz. The DFT is given by

$$X(k) = \sum_{n=0}^{N-1} x(n) \exp\left(-j\left(\frac{2\pi}{N}\right)nk\right)$$

(AVIII.1)

where:

k is the harmonic number, $0 \leq k \leq N-1$
$X(k)$ is the amount of signal at the harmonic, k
n is the sample number
$x(n)$ is the value of x at the sampling instance
N is the number of samples

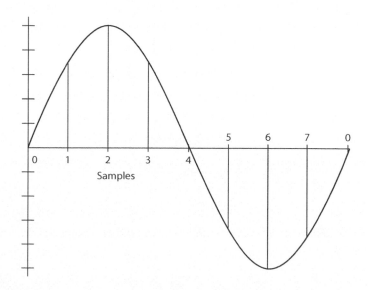

FIGURE AVIII.1   A sampled sine wave.

171

Consider $n = 0$. The magnitude of the sample $n = 0$ is zero. Hence, $x(0) = 0$. The exponential term can be split into $\cos - j\sin$ (Euler's expression). As $n = 0$, the exponential becomes $\cos 0 - j\sin 0$, which is $1 - j0$. This calculation needs to be done for all k, i.e. $k = 0$ to $N - 1$. Table AVIII.1 lists the values of $x(n)$ and $\cos$ and $j\sin$ for all n and k.

The inverse DFT is given by

$$x(n) = \sum_{n=0}^{N-1} X(k) \exp\left(+j\left(\frac{2\pi}{N}\right) nk \right)$$ (AVIII.2)

We are interested in a range of sine waves that are harmonics of a fundamental. So, consider a single sine wave as in Figure AVIII.1.

$$v_c(t) = V_c \exp\left(j(\omega_c t + \varphi_c) \right)$$ (AVIII.3)

where:

V_c is the amplitude of the sine wave (carrier)
ω_c is the angular frequency of the sine wave
φ_c is the phase of the carrier

We should note that all three parameters can be altered although it is more usual to vary the amplitude and phase of a carrier. The carrier frequency is the fundamental of a series of carriers that are harmonics of the original. So,

$$\omega_c = \omega_0 + k\Delta\omega$$

where:

ω_0 is the fundamental frequency
k is the harmonic number
$\Delta\omega$ is the spacing of the carriers

(In our example, we have a fundamental of 1 kHz and a Δf of 1 kHz with $0 \le k \le N - 1$.)

Hence,

$$v_k(t) = V_k \exp\left(j(\omega_0 + k\Delta\omega)t + \varphi_k) \right)$$ (AVIII.4)

We can arbitrarily set the starting frequency, ω_0, to zero to give

$$v_k(t) = V_k e^{jk\Delta\omega t} e^{j\varphi_k}$$ (AVIII.5)

The orthogonal frequency-division multiplexing (OFDM) signal is the sum of all of these carriers and so

$$*v_{OFDM}(t) = \sum V_k e^{jk\Delta\omega t} e^{j\varphi_k}$$ (AVIII.6)

TABLE AVIII.1
Coefficients for n (Sample Number) and k (Harmonic Number)

n	x(n)	k = 0	k = 1	k = 2	k = 3	k = 4	k = 5	k = 6	k = 7
0	0	1 − j0	1 − j0	1 − j0	1 − j0	1 − j0	1 − j0	1 − j0	1 − j0
1	0.707	1 − j0	0.707 − j0.707	0 − j1	−0.707 − j0.707	−1 − j0	−0.707 + j0.707	0 + j1	0.707 + j0.707
2	1.000	1 − j0	0 − j1	−1 + j0	0 + j1	1 + j0	0 − j1	−1 + j0	0 + j1
3	0.707	1 − j0	−0.707 − j0.707	0 + j1	0.707 − j0.707	−1 − j0	0.707 + j0.707	0 + j1	−0.707 + j0.707
4	0	1 − j0	−1 − j0	1 + j0	−1 + j0	1 + j0	−1 + j0	1 + j0	1 − j0
5	−0.707	1 − j0	−0.707 + j0.707	0 − j1	0.707 + j0.707	1 − j0	0.707 − j0.707	0 + j1	−1 + j0
6	−1	1 − j0	0 + j1	−1 + j0	0 − j1	1 + j0	0 + j1	−1 + j0	1 + j0
7	−0.707	1 − j0	0.707 + j0.707	0 + j1	−0.707 + j0.707	−1 − j0	−0.707 − j0.707	1 + j0	−1 + j0

If we sample this signal with N samples ($N = 8$ in our example) over a period T, we get

$$v_{\text{OFDM}}\left(nT\right) = \frac{1}{N}\sum_{k=0}^{N-1}V_k e^{j\varphi_k}e^{jk\Delta\omega nT} \tag{AVIII.7}$$

There are N samples per symbol time T and so,

$$\Delta f = \frac{1}{T}\frac{1}{N}$$

Therefore, Equation AVIII.7 becomes

$$v_{\text{OFDM}}\left(nT\right) = \frac{1}{N}\sum_{k=0}^{N-1}V_k e^{j\varphi_k}e^{j\left(\frac{2\pi}{N}\right)nk} \tag{AVIII.8}$$

This can be compared to the inverse DFT:

$$x\left(n\right) = \sum_{n=0}^{N-1}X\left(k\right)e^{j\left(\frac{2\pi}{N}\right)nk}$$

It can be seen that the two are equivalent meaning that the inverse DFT gives the orthogonal harmonic sine waves that we require for OFDM.

Appendix IX: Summation and Multiplication Tables in GF(8)

TABLE AIX.1
A Four-Bit Addition Table for GF(8)

	0	1	2	3	4	5	6	7
0	0	1	2	3	4	5	6	7
1	1	0	3	2	5	4	7	6
2	2	3	0	1	6	7	4	5
3	3	2	1	0	7	6	5	4
4	4	5	6	7	0	1	2	3
5	5	4	7	6	1	0	3	2
6	6	7	4	5	2	3	0	1
7	7	6	5	4	3	2	1	0

TABLE AIX.2
A Four-Bit Multiplication Table for GF(8)

	0	1	2	3	4	5	6	7
0	0	0	0	0	0	0	0	0
1	0	1	2	3	4	5	6	7
2	0	2	4	6	3	1	7	5
3	0	3	6	5	7	4	1	2
4	0	4	3	7	6	2	5	1
5	0	5	1	4	2	7	3	6
6	0	6	7	1	5	3	2	4
7	0	7	5	2	1	6	4	3

Appendix X: Bessel Function Coefficients

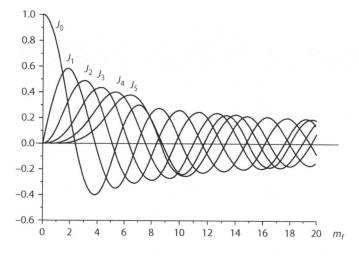

FIGURE AX.1 Plot of Bessel function coefficients.

TABLE AX.1
Listing of Bessel Function Coefficients

mf	J_0	J_1	J_2	J_3	J_4	J_5	J_6	J_7	J_8
0.0	1	0	0	0	0	0	0	0	0
0.5	0.9385	0.2423	0.0306	0	0	0	0	0	0
1.0	0.7652	0.4401	0.1149	0.0196	0	0	0	0	0
1.5	0.5118	0.5579	0.2321	0.0610	0.0118	0	0	0	0
2.0	0.2239	0.5767	0.3528	0.1289	0.0340	0	0	0	0
2.5	−0.0484	0.4971	0.4461	0.2166	0.0738	0.0195	0	0	0
3.0	−0.2601	0.3391	0.4861	0.3091	0.13200	0.0430	0.0114	0	0
3.5	−0.3801	0.1374	0.4586	0.3868	0.2044	0.0804	00254	0	0
4.0	−0.3971	−0.0660	0.3641	0.4302	0.2811	0.1321	0.0491	0.0152	0
4.5	−0.3205	−0.2311	0.2178	0.4247	0.3484	0.1947	0.0843	0.0300	0.0091
5.0	−0.1776	−0.3276	0.0466	0.3648	0.3912	0.2611	0.1310	0.0534	0.0184
6.0	0.1506	−0.2767	−0.2429	0.1148	0.3576	0.3621	0.2458	0.1296	0.0565
6.5	0.2601	−0.1538	−0.3074	−0.0353	0.2748	0.3736	0.2999	0.1801	0.0880
7.0	0.3001	−0.0047	−0.3014	−0.1676	0.1578	0.3479	0.3392	0.2336	0.1280

Appendix XI: The Phase-Lock Loop

Figure AXI.1 is a representation of a phase-lock loop (PLL). The frequency modulation (FM) is mixed with the output of a voltage-controlled oscillator (VCO) to give an error voltage. Thus,

$$v_{\mathrm{FM}}(t) = V_0 \sin\left(\omega_{\mathrm{IF}}t + \varphi_m(t)\right) \tag{AXI.1}$$

and

$$v_{\mathrm{VCO}}(t) = V_{\mathrm{VCO}} \cos\left(\omega_{\mathrm{VCO}}t + \varphi_{\mathrm{VCO}}(t)\right) \tag{AXI.2}$$

where V_0 and V_{VCO} are the amplitudes of the incoming FM signal and the VCO output. The FM is deliberately amplitude limited to remove any amplitude modulation (AM) on the signal. Note that the VCO term is a cosine – a phase shift of 90°. The reason for this will become clear soon. The loop is not locked as shown by ω_{IF} and ω_{VCO} being different. The phase shift φ_m is related to the modulation by

$$\varphi_m(t) = 2\pi k_f \int V_m \sin \omega_m t \tag{AXI.3}$$

This is the same as in the main text. Similarly,

$$\varphi_{\mathrm{VCO}}(t) = 2\pi k_{\mathrm{VCO}} \int v_{\mathrm{error}}(t) \tag{AXI.4}$$

The parameters k_f and k_{VCO} are the constants of proportionality for the VCO in the FM modulator and the demodulating PLL.

The product of the FM and the VCO is

$$v_{\mathrm{FM}}(t)v_{\mathrm{VCO}}(t) = V_0 V_{\mathrm{VCO}} \sin\left(\omega_{\mathrm{IF}}t + \varphi_m(t)\right)\cos\left(\omega_{\mathrm{VCO}}t + \varphi_{\mathrm{VCO}}(t)\right)$$

$$= \frac{V_0 V_{\mathrm{VCO}}}{2}\left\{ \begin{array}{l} \sin\left(\omega_{\mathrm{IF}}t + \omega_{\mathrm{VCO}}t + \varphi_m(t) + \varphi_{\mathrm{VCO}}(t)\right) \\ + \sin\left(\omega_{\mathrm{IF}}t - \omega_{\mathrm{VCO}}t + \varphi_m(t) - \varphi_{\mathrm{VCO}}(t)\right) \end{array} \right\} \tag{AXI.5}$$

The mixer is followed by a low-pass filter which will filter out the high-frequency first term in Equation AXI.5. Therefore, the error voltage will be

$$v_{\mathrm{error}}(t) = A_v \frac{V_0 V_{\mathrm{VCO}}}{2} \sin\left(\omega_{\mathrm{IF}}t - \omega_{\mathrm{VCO}}t + \varphi_m(t) - \varphi_{\mathrm{VCO}}(t)\right) \tag{AXI.6}$$

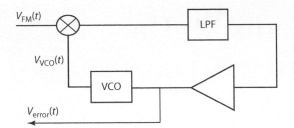

FIGURE AXI.1 Block diagram of a phase-lock loop.

where A_v is the gain of the amplifier. The loop will try to lock to the intermediate frequency (IF) of the FM signal by producing the error voltage which acts to lock the VCO to the IF. So, Equation AXI.6 becomes

$$v_{\text{error}}(t) = A_v \frac{V_0 V_{\text{VCO}}}{2} \sin\left(\varphi_m(t) - \varphi_{\text{VCO}}(t)\right) \quad \text{(AXI.7)}$$

As the loop attains lock, $\varphi_m(t) - \varphi_{\text{VCO}}(t)$ tends to zero and so $\sin(\varphi_m(t) - \varphi_{\text{VCO}}(t))$ tends to $\varphi_m(t) - \varphi_{\text{VCO}}(t)$. Thus, the error voltage becomes

$$v_{\text{error}}(t) = A_v \frac{V_0 V_{\text{VCO}}}{2}\left(\varphi_m(t) - \varphi_{\text{VCO}}(t)\right)$$

Therefore,

$$v_{\text{error}}(t) \frac{2}{A_v V_0 V_{\text{VCO}}} = \varphi_m(t) - \varphi_{\text{VCO}}(t) \quad \text{(AXI.8)}$$

If the gain of the amplifier is large, $2/2\pi A_v V_0 V_{\text{VCO}}$ tends to zero. Hence, Equation AXI.8 becomes

$$\varphi_{\text{VCO}}(t) \approx \varphi_m(t) \quad \text{(AXI.9)}$$

And so,

$$2\pi k_{\text{VCO}} \int v_{\text{error}}(t) = 2\pi k_f \int V_m \sin \omega_m t$$

Differentiating both sides gives

$$2\pi k_{\text{VCO}} v_{\text{error}}(t) = 2\pi k_f V_m \sin \omega_m t$$

Hence,

$$v_{\text{error}}(t) = \frac{k_f}{k_{\text{VCO}}} V_m \sin \omega_m t \quad \text{(AXI.10)}$$

So, the error voltage that is applied to the VCO to maintain lock as the FM signal varies in frequency is directly proportional to the audio signal. The PLL has demodulated the FM.

Appendix XII: Lumped Parameters for Coaxial Cable

Figure AXII.1 shows a cross section through a piece of coaxial cable. The inner conductor has a radius a and the outer shield has a radius b. We will take a thin tube of radius r and thickness dr and calculate the capacitance, inductance and conductance.

Electric flux will emanate from the inner conductor by virtue of it being at a higher voltage than the outer. This will give capacitance. The flux will be Q Coulombs and so the flux density, D, at radius r will be

$$D_r = \frac{Q}{2\pi r \times l} \, \text{C/m}^2$$

Thus, the electric field strength at radius r is

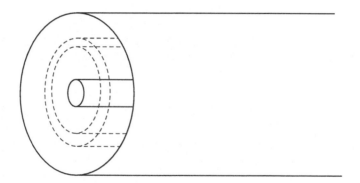

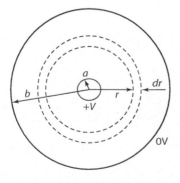

FIGURE AXII.1   (a) Section through a length of coaxial cable and (b) showing the definitions of various radii.

181

$$E_r = \frac{Q}{2\pi\varepsilon_0\varepsilon_r r \times l}\ \text{V/m}$$

where:

   ε_0 is the permittivity of free space

   ε_r is the relative permittivity of the dielectric used in the cable

This electric field will give a potential across the tube of dV. So,

$$dV = -E_r dr$$

Therefore

$$\int_0^V dV = -\int_b^a \frac{Q}{2\pi\varepsilon_0\varepsilon_r r \times l}\, dr\ \text{V}$$

And so

$$V = \frac{Q}{2\pi\varepsilon_0\varepsilon_r \times l}\ln\left(\frac{b}{a}\right)\text{V}$$

As $Q = CV$:

$$C = \frac{2\pi\varepsilon_0\varepsilon_r \times l}{\ln\left(\dfrac{b}{a}\right)}\ \text{F}$$

$$C = \frac{2\pi\varepsilon_0\varepsilon_r}{\ln\left(\dfrac{b}{a}\right)}\ \text{F/m} \qquad\qquad (\text{AXII.1})$$

As regards the shunt resistance, the conductance, it is current that flows from the centre conductor to the shield. So, the current density at radius r is

$$J_r = \frac{I}{2\pi r \times l}\ \text{A/m}^2$$

As $J = \sigma E$:

$$E_r = \frac{I}{2\pi\sigma r \times l}\ \text{V/m}$$

where σ is the conductivity of the dielectric. This electric field will give a potential across the tube of dV. So,

$$dV = -E_r dr$$

Therefore

$$\int_0^V dV = -\int_b^a \frac{I}{2\pi\sigma r \times l} \, dr \; \text{V}$$

And so

$$V = \frac{I}{2\pi\sigma \times l} \ln\left(\frac{b}{a}\right) \text{V}$$

As $V = IR$:

$$R_{\text{shunt}} = \frac{1}{2\pi\sigma \times l} \ln\left(\frac{b}{a}\right) \text{V} \qquad\qquad (\text{AXII.2})$$

There are two parts to the inductance: the inductance due to the magnetic field inside the inner conductor and the field in the dielectric. The inner inductance is given by

$$\frac{\mu_0}{8\pi} \; \text{H/m} \qquad\qquad (\text{AXII.3})$$

This represents a fixed inductance and, as μ_0 is $4\pi \times 10^{-7}$ H/m, it is 50 nH/m. This is significant at high frequencies. As regards the external inductance, the current in the inner conductor generates a magnetic field in the thin tube of

$$H = \frac{I}{2\pi r} \; \text{A/m}$$

As $B = \mu_0 H$:

$$B = \frac{\mu_0 I}{2\pi r} \; \text{Wb/m}$$

Therefore, the flux in the tube is

$$d\Phi = \frac{\mu_0 I}{2\pi r} \, dr \, \text{length Wb}$$

The fractional inductance per unit length is

$$\frac{d\Phi}{I} = \frac{\mu_0}{2\pi r} \, dr \; \text{H/m}$$

And so the total inductance is

$$L = \frac{\mu_0}{2\pi} \int_a^b \frac{1}{r} \, dr \; \text{H/m}$$

$$L = \frac{\mu_0}{2\pi} \ln\left(\frac{b}{a}\right) \text{ H/m} \qquad\qquad\qquad \text{(AXII.4)}$$

The resistance of the inner conductor is simply

$$R = \frac{\rho \text{ length}}{\pi a^2} \, \Omega \qquad\qquad\qquad\qquad \text{(AXII.5)}$$

Appendix XIII: The 4B5B Line Code

Incoming Data	Coded Data
0000	11110
0001	01001
0010	10100
0011	10101
0100	01010
0101	01011
0110	01110
0111	01111
1000	10010
1001	10011
1010	10110
1011	10111
1100	11010
1101	11011
1110	11100
1111	11101

Index